D1414297

About Island Press

Island Press, a nonprofit organization, publishes, markets, and distributes the most advanced thinking on the conservation of our natural resources — books about soil, land, water, forests, wildlife, and hazardous and toxic wastes. These books are practical tools used by public officials, business and industry leaders, natural resource managers, and concerned citizens working to solve both local and global resource problems.

Founded in 1978, Island Press reorganized in 1984 to meet the increasing demand for substantive books on all resource-related issues. Island Press publishes and distributes under its own imprint and offers these services to other nonprofit organizations.

Support for Island Press is provided by Apple Computers, Inc., Mary Reynolds Babcock Foundation, Geraldine R. Dodge Foundation, The Charles Engelhard Foundation, The Ford Foundation, Glen Eagles Foundation, The George Gund Foundation, William and Flora Hewlett Foundation, The Joyce Foundation, The John D. and Catherine T. MacArthur Foundation, The Andrew W. Mellon Foundation, The Joyce Mertz-Gilmore Foundation, The New-Land Foundation, The J. N. Pew, Jr. Charitable Trust, Alida Rockefeller, The Rockefeller Brothers Fund, The Florence and John Schumann Foundation, The Tides Foundation, and individual donors.

About the Environmental Action Coalition

Founded in 1970, the Environmental Action Coalition (EAC) is a not-for-profit corporation specializing in environmental education and in active projects that involve individuals on the grassroots level in processes of protecting and enhancing environmental quality. EAC has focused the bulk of its work in the New York City metropolitan region, but has increasingly become active throughout New York State and the Middle Atlantic region. In addition, through links with other organizations and research projects, the Coalition has influence on national policy and on the fostering of broad-based educational outreach. Many EAC staff sit on local boards and advisory councils in which they are able to work closely with individuals and community groups.

EAC focuses on recycling, providing the organization and financing for New York City's first voluntary recycling network in the early 1970s. Since that time, staff and board have worked for the establishment of an integrated waste management program, urging public/private cooperative measures. Recent programs include apartment house recycling, institutional office paper recycling, battery recycling, and household hazardous waste management. EAC's Environmental Education Program developed some of the nation's first urban environmental education curriculum supplements on a variety of topics for young students. EAC's in-school programs center on recycling education, water conservation, and urban ecology. EAC also promotes the planting of city trees, and currently functions as part of the New York State coordination team for New York ReLeaf, a comprehensive tree-planting program. The Coalition provides an extensive national information service on all environmental issues. EAC's library, one of the few comprehensive environmental libraries in the country, and its publications are available to the general public. Membership is available by writing to 625 Broadway, New York, NY 10012; (212) 677-1601.

PLASTICS

PLASTICS

America's Packaging Dilemma

By
Nancy Wolf and
Ellen Feldman

ENVIRONMENTAL ACTION COALITION

ISLAND PRESS

Washington, D.C. • Covelo, California

Library of Congress Cataloging-in-Publication Data

Wolf, Nancy A.
    Plastics: America's packaging dilemma / by Nancy A. Wolf and Ellen D. Feldman.
        p.   cm. — (Island Press critical issues series; #3)
    Includes bibliographical references and index.
    ISBN 1-55963-063-9. — ISBN 1-55963-062-0 (pbk.)
    1. Plastic scrap—Environmental aspects—United States. 2. Plastics in packaging—United States. I. Feldman, Ellen D. II. Title. III. Series.
TD798.W64   1990
363.72'88—dc20   90-43801
                                                                      CIP

Printed on recycled acid-free paper

Manufactured in the United States of America
10 9 8 7 6 5 4 3 2 1

Contents

Abbreviations

EVA	ethylene vinyl acetate
EVOH	ethylene vinyl alcohol
HDPE	high-density polyethylene
LDPE	low-density polyethylene
LLDPE	linear low-density polyethylene
PCB	polychlorinated biphenyl
PCDD	polychlorinated dibenzo-p-dioxin
PCDF	polychlorinated dibensofuran
PET	polyethylene terephthalate
PIC	product of incomplete combustion
PP	polypropylene
PS	polystyrene
PVC	polyvinyl chloride
PVDC	polyvinylidene chloride

Preface

Plastics: America's Packaging Dilemma is the most serious and potentially significant research yet undertaken by the Plastics Research Project of the Environmental Action Coalition (EAC). Funded by the Robert Sterling Clark Foundation and the New York Community Trust, EAC Staff were able to conduct a year-long investigation into the current state of plastics use and design, methods for disposal, and trends that will impact recyclability and the management of solid waste. The work done by Ellen Feldman, Research Associate, is built upon previous research conducted by George Pess of Bowdoin College and Angela Lui of Brown University, who both worked at EAC under the sponsorship of the Environmental Intern Program.

EAC's research on plastic packaging and its recyclability began in the spring of 1984 and was prompted by two major developments.

1. Local and state governments increasingly had begun to adopt source-separation recycling of municipal solid waste as a major part of their waste disposal strategies.

2. Although it was commonly perceived that many traditional recyclables, such as metals, glass, and cardboard, had been replaced by differing types of plastics, no current and relevant data then existed that would verify percentages or totals to guide projections of the recyclability of the waste stream.

Many working in the recycling field were concerned that erroneous assumptions about the recyclability of the waste stream were being made, using statistics gathered before the rapid proliferation of plastic packaging began. It was feared that the use of such outdated figures would inflate estimates of how recycling could alleviate the increasing solid-waste crisis. Could recycling take 25 percent to 50 percent of the waste stream, if the very materials being relied on for those estimates were being replaced by plastics of uncertain recyclability?

Thorough examination of this development had not yet taken place, and its potential seriousness was just being recognized. Due in part to EAC's continued active recycling projects and the research associated with them, many facts surrounding the changes in the waste stream and its present and potential recyclability are being made known. As public-policy questions are examined and continuing executive and legislative actions taken, it is expected that the information in the following study will be of assistance to all involved in the investigation of what to do with garbage.

Although based on technological and scientific secondary research and certain empirical data-gathering, this report is not designed primarily for the small number of active professionals now in the field of plastics research. Rather, as the issues surrounding plastics become ever more controversial, this factual report will be of use to lay readers in frontline decisionmaking roles: legislators and executives at all levels of government, planners from many disciplines, and active citizen leaders.

Caution must be observed, however, in terms of many statistics gathered and quoted in this book. The secondary sources cited come almost exclusively from industry-derived figures and could not be verified independently. Many, no doubt, have changed considerably since their original publication. Primary data has been gathered largely through examination of EAC's own recycling projects and through special surveys in supermarkets and drugstores that were organized by EAC staff.

EAC would like to thank its two peer reviewers—Gretchen Brewer, private consultant previously with the State of Massachusetts and Marjorie Clarke, researcher with INFORM. They were kind enough to read the early draft version and their corrections, comments, and suggestions were invaluable. It is hoped that this final document will be both accurate and readable.

May 1990

In November 1990 as this book was going to press, the McDonald's Corporation announced its abandonment of the polystyrene foam "clamshell" package. The decision, which was arrived at by top management, took most of the country by surprise since McDonald's had been expected to announce the expansion of its pilot recycling program in the northeast to many other areas of the country. Ironically, McDonald's decision to switch to bleached paper/polyetheylene wrappers led the company to abandon a recyclable package in favor of one that is neither recyclable nor compostable.

In an interview with Environmental Action Coalition staff on November 5, McDonald's Senior Vice President, Shelby Yastrow, attributed the decision more to economic factors than to environmental ones. The lack of commitment from their polystyrene-resin producing partners in the National Polystyrene Recycling Corporation had created a financial drain on the company. Additionally, despite great efforts, McDonald's could not convince customers that the foam box—made without CFCs as a blowing agent—was an environmentally desirable package.

What is next for environmentalists and concerned consumers? Logically, since McDonald's and partners represented the best hope of a nationwide PS foam recycling infrastructure, perhaps the question of banning all polystyrene foam in the retail marketplace ought to be seriously examined. In any event, the backing down by a major company, due to citizen pressure— however wrongly informed—shows the power of the consumer, in the end.

November 1990

PLASTICS

Executive Summary

Plastics constitute one of the fastest-growing categories of materials used — and disposed of — in our economy. Today, plastics comprise about 8 percent of the weight and nearly 30 percent of the volume of the municipal solid-waste stream. Projections for the year 2000 estimate that plastic materials will account for at least 10 percent by weight of the local solid-waste stream.

Recycling processes, now becoming a key strategy for solid-waste disposal, are firmly established for materials such as glass, paper, and metals, but the recycling of plastics is in its infancy.

Currently, the two largest market areas for plastics in the United States are packaging and the building and construction industries. In both categories, annual usage roughly doubled between 1974 and 1984. If current projections prove accurate, in 1995 the United States will consume 19 billion pounds of plastic packaging and 14.5 billion pounds of plastic building and construction materials.

Plastic packaging materials have a life span of less than one year. These materials quickly enter the waste stream, with each pound of plastic packaging producing just under a pound of waste. In contrast, plastics used in building and construction generally have a life span of 25 to 50 years. Substantial amounts of plastic wastes from construction completed during the 1960s and 1970s are expected to enter the waste stream in the late 1990s and into the next century.

Plastic packaging materials are composed of a variety of different resins and resin combinations. The most common are polyethylene terephthalate (PET), high-density polyethylene (HDPE), low-density polyethylene (LDPE), polyvinyl chloride (PVC), polypropylene (PP), and polystyrene (PS). The two largest categories of current use are rigid containers and packaging film, constituting 51 percent and 31 percent of plastics in packaging, respectively.

Other synthetics used in packaging include ethylene vinyl acetate (EVA) copolymer, polyvinyl acetate as an adhesive, and ethylene vinyl alcohol (EVOH) as an oxygen barrier. The use of these materials reflects an important recent development in plastic packaging—the use of composite, or multilayer, packaging. Since the composites contain many different types of resins and other materials in many layers, they are extremely difficult, if not impossible, to recycle.

Many of the composite packages are replacing packaging materials with high recycling histories, such as glass and paperboard. The disposability of composite packages and the environmental impacts of disposal have been largely ignored in decision making. As these complex combinations of plastic become more common, the recyclability potential of the waste stream will decrease, since there is no recycling mechanism for these combination packages.

The growth in the use of plastics for building and construction is projected to be large through the year 2000. There are relatively few wastes from these categories in the present waste stream, since only small amounts of plastics were used in construction during the 1950s. But the "revolution" in the use of plastics for building and construction that began in the 1960s will impact the waste stream greatly in the 1990s and in the next century.

Consumer and institutional goods represent the third largest use of plastics. This category includes all carryout packages; disposable plastic serviceware in institutions such as schools, hospitals, and prisons; and throwaway razors, lighters, pens, watches, and cameras. Other areas of plastics use include electrical and electronic equipment, furniture and furnishings, and industrial applications.

Food Packaging

In judging the impact of plastic on the waste stream, particularly in landfills, volume, not weight, is the relevant measure. Because plastics are so light in comparison to their volume, they occupy a much larger space relative to weight than do most other materials. The substitution rate of plastics for older food-packaging materials is significant, especially in the soft-drink sector. New package design, particularly barrier packaging that prevents leaching of packaging materials into food, has allowed the proliferation of plastic packages for new applications. Plastic is often the material of choice because it is lightweight, microwaveable, durable, and resistant to breakage. Thus flexible packaging is rapidly replacing paper in the grocery-sack market and foil in the dual-ovenable market. But package designers' areas of expertise are engineering, design, and marketing, not environmental protection. Therefore, many "exciting" new packages rapidly become disposal problems once their useful lives are complete, being neither recyclable nor naturally biodegradable.

Some package designers are gradually becoming aware of their role in the problems of waste disposal. Designers now working in the field are reeducating themselves and their colleagues, and environmental design is becoming part of the coursework in design schools. Environmentalists are seeking to influence this emerging trend, hoping to add disposal considerations to food packagers' concerns about marketing, cost, shelf life, and competition for shelf space.

The growing controversy over packaging materials in the United States has caught many corporations by surprise and has forced a reevaluation of packaging decisions at the top of the decision-making ladder. Public pressure is beginning to have some effect. Among the food-processing giants seeking to come to grips with the new consumer and environmental demands are Campbell, Hormel, and Best Foods. The decisions that will be made by these leading companies may be expected to set the agenda for the rest of the industry.

In some cases, consumer groups have singled out companies for special targeting—the case of McDonald's, still defending its use of polystyrene "clamshells," is especially visible. Although polystyrene is technically recyclable, the company has only recently promised to put in place a national recycling infra-structure for collection and recycling from all its outlets. While the company is now sponsoring pilot projects involving a few outlets, public pressure has increased, rather than abated. Another high-visibility controversy concerns the increasing use of plastic film for grocery bags, replacing recyclable and compost-able paper bags. In debate over plastics versus paper, unfortu-nately, most of the research studies are somewhat suspect, hav-ing been funded by the industries involved.

Any change back to traditional recyclable packages such as paper, paperboard, glass, and metals now seems extremely dif-ficult, given the years of decision making in the other direction. A survey conducted for the Environmental Action Coalition (EAC) in spring 1988 found that over half of all supermarket items surveyed were packaged entirely in plastic, either single layer or multiple layered. In some categories nearly all items were packaged in plastic: dairy, produce, meat, and household cleaning items. A survey in drugstore chains found that over 75 percent of products surveyed had some plastic in their pack-aging, with 40 percent in all-plastic packaging. Only 1.3 percent of the packaging surveyed in drugstores was all paper; 3.9 per-cent had all or some glass; 12.1 percent was all cardboard.

A survey of participants in New York City's new curbside recycling program reported that households found relatively few items available for inclusion, other than wine bottles. House-holds were buying the same basic products as always, but, increasingly, those products are packaged in containers not part of the recycling program. Without changes in packaging design and an increase in the recyclability of packages, it may be extremely difficult for New York City and other localities to meet the recycling targets they have set in law.

Spurred on by the large resin producers, such as Mobil, DuPont, and Dow, package designers are paying more attention to packaging made from only one type of resin. Statistics show that these packages have the highest chance of expanded recyclability. An example of such a change is the development of an all-PET beverage container that does away with the basecup made of HDPE. Although HDPE is itself recyclable, the need to separate the HDPE basecup from the PET body made the original plastic bottle more expensive to recycle than competing products. Pushed by the same emerging consortium of companies, the industry is now promoting voluntary coding by which consumers and recyclers can clearly identify, separate, and collect recyclable plastics. While resistance to coding and arguments over whether or not the industry has the right to incorporate in its code the circle of recycling arrows are slowing the adoption of coding, this new understanding will help facilitate higher recycling volumes of at least some packaging.

Environmental Impacts

The increases in such disposables as six-pack beverage container rings, tampon applicators, fast-food packaging, and diapers has alarmed most conservationists. The presence of waste plastics in waterways has been especially harmful and has led to federal action through the Marine Plastic Pollution Research and Control Act of 1987. Plastic debris from boats or from beach litter has been shown to pose serious threats to nearby marine animals, via either entanglement or ingestion.

On land, there are many more potential problems in routine,

daily solid-waste disposal by landfilling or incineration. In land-fills, plastics, because of their synthetic composition and inert qualities, provide increasing volumes of materials that do not break down. Furthermore, landfills may be contaminated by the possible leaching of additives and stabilizers in plastics. While inert plastics may not cause immediate problems in landfills being managed for maximum compaction of all garbage and trash, they clearly hinder any management of landfills for com-posting, degradation of contents, and recovery of naturally oc-curring methane gas.

One of the most intense controversies concerns the presence of various plastics in resource-recovery incineration. Industry representatives tout the high BTU content of plastics and see incineration as desirable for disposal of these wastes. But some environmentalists claim to have linked the combustion of plas-tics in mixed wastes to the production of dangerous pollutants. In studies reviewed by EAC, only one clear link was found between incineration of plastics and recognized pollutants: PVC in incineration batches creates corrosion and increased hydro-chloric acids. Much more work remains to be done on this subject, as with the study of combinations of plastics in land-fills.

Recycling

Recycling of plastics, like other materials, is highly desirable based on energy use alone. In all cases, the use of process energy decreases with the remanufacture of items from already pro-duced resources. The recycling of scrap plastics from the pro-duction line has been in place for many years.

However, to date there has been very little recycling of postconsumer plastics, with the exception of the emerging infra-structure to support the collection and recycling of PET beverage containers from "bottle bill" states. In contrast to the aluminum industry, where support of the recycling infrastructure has long been in place, recycling of plastics accounts for only about 1 percent of all products.

The recycling of plastics is different from the recycling of glass and metals. Whereas glass bottles and metal cans come back into use in the same mode, many recycled plastics must be made into other products, due to the inability of plastics to be remanufactured and sterilized to meet food-contact standards. While some pilot programs are beginning to produce recyclable containers for nonfood uses, most recycled plastics become fiber that is used in such products as carpets or jacket filler. Combined plastics are beginning to be made into "plastic lumber" products for use in such applications as park benches.

Most efforts, to date, have been concentrated on the reclamation of PET and HDPE containers, representing 7 percent and 26 percent of the plastic packaging waste stream, respectively. The "bottle bill" collections provide a good source of product for PET recycling, while the HDPE collections are more diverse, primarily undertaken by scattered community recycling programs. As coding becomes more widespread, these two resins (if not contaminated by being combined with other plastics or with other materials) could be phased into larger recycling programs, such as curbside pickup. As with the recycling of other materials, however, much public education is needed, including identification methods, preparation requirements, and collection frequency and form.

One of the major problems in the recycling of postconsumer plastics is the high cost of collection and transportation. Most empty packages occupy a large volume: transporting them has been likened to carrying around inflated balloons. Small towns may not be able to collect enough tonnage of plastics to sustain their recyclability as part of an economically viable program. One proposed solution calls for plastics manufacturers to assist localities in financing grinders and other equipment that reduce plastics into a denser form that is more economical to transport.

Despite these difficulties, pilot programs for recycling plastics exist and more will be developed. It is much too soon to predict how effective they will be.

Legislative Efforts

Today's national controversies over plastics will have an enormous effect on the myriad local and state bodies whose decisions govern the production, use, and disposal of plastics. The industry, which for many years was viewed in positive ways by most, is now the target of increasing hostility and consumer negativity. While much of the opposition is based on emotion and often faulty information, there can be no doubt that the honeymoon of plastics and the consumer is coming to an end.

One major debate is over the "degradability" of plastics. This is one of the most disputed topics and one in which the industry has itself become a major culprit. In an effort to respond to local and state laws mandating an ill-defined "degradability," the plastics industry is now claiming "biodegradability" and "photodegradability" for various products, especially six-pack loops and plastic bags. Much more research needs to be done, but clearly true degradability is not a property of synthetic polymers and many of the companies' claims to "degradability" are primarily public relations strategies.

True biodegradability is a natural function and connotes the return of a material to the soil base. Synthetic plastics are not, and cannot be, biodegradable because they are inert. In order to make plastics degradable, producers mix natural compounds, like starch, into the inert plastic film. When the starch breaks down in the presence of oxygen and moisture, the plastic polymer is left behind in the environment—as shards or "dust."

"Photodegradability" has been achieved, according to the plastics industry, by adding various percentages of chemical additives—which are protected by proprietary information, in most cases—to the polymers. In reaction to ultraviolet light, these chemicals break down into the environment, again leaving shards or "dust" plastics behind.

The industry's claim to having produced so-called degradable plastics is one of the most alarming current trends. Set off by ill-defined laws, this situation has led to a proliferation of contradictory laws at local and state levels and an increase in the production of items that often mislead the inexperienced consumer. At worst, chemicals added for "photodegradation" can be harmful to the surrounding environment if they break down.

The slow response of the plastics industry to municipal waste-disposal problems has given rise to proposals for a myriad of local and state bills to restrict the use of plastics. Some of these have become law. In the opinions of most of those active in the field, these initiatives were a response to real or perceived crises that were being ignored by the federal government, where, perhaps, solutions to the controversies lie.

Among the proposed or already enacted restrictive measures are returnable beverage container laws (commonly known as "bottle bills"), packaging taxes based on the recyclability of a product or its use of recycled material, and even certain bans, such as the now-famous partial ban on plastics use enacted by Suffolk County, New York. Most environmentalists consider the explosion of bills and potential laws to be inevitable, considering the inertia evident at the national level. In light of the wide diversity of proposed local initiatives, the industry is now joining many environmentalists in calling for national action by the Congress or the U.S. Environmental Protection Agency.

The answers to five critical questions will determine how the U.S. copes with the growing seas of plastic:

- Will the federal government analyze the nationwide production of packaging and products and recommend reduction, both in toxicity and in volume?
- Will the federal government analyze the plethora of local and state measures and organize them into coherence?

- Will local, state, and national government come to terms with the institutional barriers that restrict the success of recycling—such as fire codes that prevent the use of physical spaces for collections, procurement laws that forbid or impede the purchase of goods made of recycled materials, and freight rates that favor virgin over recycled materials?
- Will the issue of mandated "biodegradability" and "photodegradability" be opened up for reexamination, leading to accepted legal definitions of those terms?
- Will a data-base of scientific and social statistics be available for informed decision making?

When Environmental Action Coalition embarked on its packaging review in 1987, its major question was, Is the recyclability of the waste stream at risk? The answer to that is now clear—an unequivocal yes. The solutions, however, are not yet in sight.

1

Uses of Plastics in the United States Economy

Plastics constitute one of the fastest-growing categories of materials used — and disposed of — in our economy. In 1970 plastics comprised less than 3 percent of the municipal solid-waste stream by weight; in 1986 the figure had risen to 6 percent (see figure 1); recent figures indicate that plastic waste comprises about 8 percent of the stream by weight and nearly 30 percent by volume.[1] Projections for the year 2000 estimate that plastics will comprise at least 10 percent of the stream by weight. However, due to new efforts to reduce packaging size, volume may increase more slowly.

The increase in plastics use, which is a national trend, may pose serious problems to waste-disposal operations, almost all of which are managed by local governments. As landfill space — long the most common disposal method — diminishes, and new landfills are becoming nearly impossible to site, solid-waste planners must look to other methods of waste management. Resource recovery through mass-burn incineration, the alternative most frequently mentioned, is often cited as an acceptable method of "recycling" waste plastic (due to the high energy content of plastics), but serious concerns over health issues have been raised. Recycling, more and more considered the most acceptable method of disposal, is firmly established for materials

FIGURE 1
Composition of Solid Waste in the United States by Weight 1986

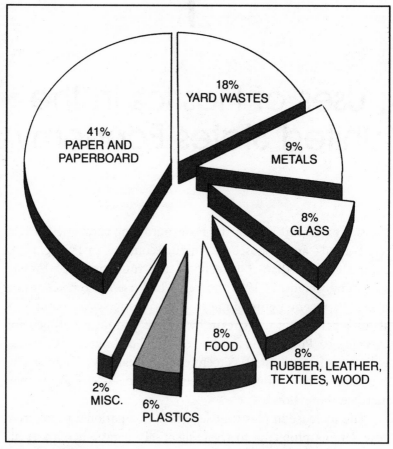

Source: Franklin Associates, Ltd. for the U.S. Environmental Protection Agency: *Characteristics of Municipal Solid Waste in the United States, 1960–2000 (Update, 1988).*

such as glass, paper, and metals, but is in its infancy in regard to plastics.

Currently, the two largest market areas for plastics in the United States are packaging and the building and construction industries. The use of plastics in packaging increased from 6.7 billion pounds in 1974 to 12.4 billion pounds in 1984. Projections indicate that use will increase to 16.1 billion pounds in 1990 and to 19.1 billion pounds by 1995.[2] Because plastic packaging materials have a life span of less than one year, consumption by the industry for this purpose is nearly equal to the amount of waste. Approximately 29.4 billion pounds of postconsumer plastic wastes were discarded in 1984; over one-third of this came from packaging that was quickly discarded.[3]

The use of plastics in building and construction materials increased from 4.3 billion pounds in 1974 to 9.7 billion pounds in 1984. Use is projected to increase to 12 billion pounds in 1990 and to 14.5 billion pounds in 1995. Since these materials generally have a life span of 25 to 50 years, substantial amounts of plastic wastes from building and construction undertaken in the 1960s and 1970s are expected to enter the waste stream in the late 1990s and into the next century.[4] If plastic becomes more widely used for these applications, disposal problems may accelerate far into the future. Thus, planning should begin now to address this situation.

Consumer and institutional goods represent the third largest use of plastics. This category includes all carryout packages; disposable plastic serviceware in institutions such as schools, hospitals, and prisons; and throwaways such as razors, lighters, pens, watches, and cameras. Other areas of use include electrical and electronic equipment, furniture and furnishings, transportation, and industrial applications.[5]

The projected growth in all categories of postconsumer plastic wastes in the United States is shown in table 1.

Packaging

It is estimated that packaging materials account for more than one-third of municipal waste, by volume, in the United States.

TABLE 1
Annual Quantities of Plastic Waste

Category	1984 Billion lbs.	% of total	1995 (projected) Billion lbs.	% of total
Packaging	12.4	42.2	19.1	44.5
Consumer goods	3.6	12.3	4.3	10.1
Construction	0.7	2.4	3.8	8.9
Furniture	2.2	7.4	3.1	7.2
Electric/ electronic	2.3	8.0	2.7	6.4
Transportation	1.9	6.6	2.3	5.2
Industrial machinery	0.4	1.2	0.43	1.0
Other	5.9	20.0	7.1	16.6

Source: Curlee, The Economic Feasibility of Recycling, pp. 82, 86.

By weight, the percentages of packaging materials in the municipal waste stream in 1986 were: 14.5 percent paper, 7.6 percent glass, 1 percent aluminum, 4 percent plastic, and 2 percent steel.[6]

The packaging industry is the largest consumer of plastics, accounting for over one-third of all plastic resin use annually. Plastics are used as films for flexible packages, for barrier material on bottles and rigid containers, for soft-drink containers, and for coating on many other items. Packaging is also the largest single source of plastic waste, reaching approximately 13 billion pounds a year, or 40 percent of all plastic waste. Most of the waste plastics are thermoplastics (see Appendix A), and waste plastics are expected to increase as packaging manufacturers continue to switch from other packaging materials. More new materials are developed in this area of the economy than in any other area.[7] The two largest categories of current use are rigid containers and packaging film, accounting for 51 percent and 35 percent of plastics, respectively (see figure 2).[8]

Plastic packaging materials are composed of a variety of different resins and resin combinations. (See Appendixes A

FIGURE 2
Uses of Plastics in Packaging 1987

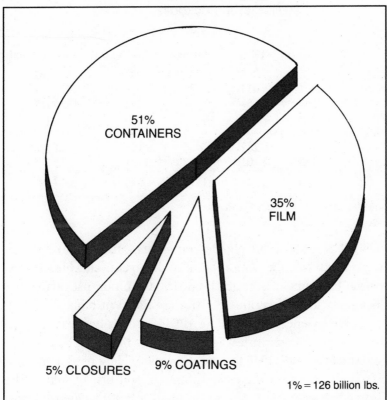

51%
CONTAINERS

35%
FILM

5% CLOSURES   9% COATINGS

1% = 126 billion lbs.

Source: Modern Plastics, January 1988.

through D.) The most common are low-density polyethylene (LDPE), high-density polyethylene (HDPE), polystyrene (PS), polyvinyl chloride (PVC), polyethylene terephthalate (PET), and polypropylene (PP). The amounts of plastics used in packaging in 1987 are summarized in table 2 and illustrated in figure 3.

Polyethylene (HDPE and LDPE) makes up over 60 percent of the plastic packaging waste stream. LDPE packaging film, the most prevalent plastic film, is used in applications such as grocery sacks and bread wrap. HDPE containers, which comprise

TABLE 2
Plastics in Packaging—1987
(in billion lbs.)

	Containers	Film	Coatings	Closures	Total (% of total)
LDPE	256	3085	742	33	4116 (32%)
HDPE	3331	414	42	89	3876 (31%)
PS	1055	57	–	234	1346 (11%)
PP	438	537	27	263	1265 (10%)
PET	900	–	12	–	912 (7%)
PVC	310	232	33	36	611 (5%)
Other	143	101	258	21	523 (4%)
Total	6433	4426	1114	676	12,649

Source: Modern Plastics, January 1988.

over 50 percent of all plastic containers, are used to package items such as milk, water, laundry detergent, and bleach, and for the basecups of soft-drink bottles. Common packaging and disposable uses of various plastics are listed in table 3.

Polyethylene terephthalate (PET), primarily used in soft-drink bottles, constitutes approximately 14 percent of plastic containers overall. PET is now the most widely used soft-drink container, based on the total volume of soft drinks sold in the United States in 1986: 43 percent of all soft drinks are packaged in PET bottles, compared to 34 percent in 12-ounce cans. In 1988 an estimated 7.5 billion PET bottles were discarded after use in the U.S.; usage is anticipated to double by 1995 to 15 billion bottles a year, or over 2 billion pounds of resin.[9]

Other materials used in packaging, to a lesser extent, include ethylene vinyl acetate (EVA) copolymer, polyvinyl acetate as an adhesive, and ethylene vinyl alcohol (EVOH) as an oxygen barrier. The use of these materials reflects an important recent development in plastic packaging—the use of composite, or multilayer, packaging. These packages contain layers—sometimes as many as 12—of different types of resins and other materials. For example, the squeezable ketchup bottle, now re-

FIGURE 3
Types of Plastics in Packaging 1987

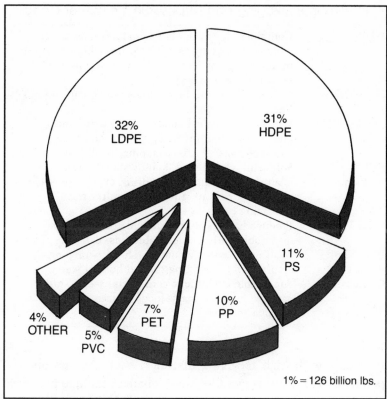

32%
LDPE

31%
HDPE

11%
PS

4%
OTHER

5%
PVC

7%
PET

10%
PP

1% = 126 billion lbs.

Source: Modern Plastics, January 1988.

placing glass, consists of a layer of polypropylene (PP), an adhesive layer, an oxygen barrier layer (EVOH), another adhesive layer, and another layer of PP. The use of EVOH as an oxygen barrier allows manufacturers to package in plastic many foods they previously could not, because of possible contamination. Another example of a composite package now in wider use is the aseptic juice box, or "brick-pack." The package combines plastic and adhesive layers with paperboard and metal foil.

Many such composite packages are now replacing packaging

TABLE 3
Plastics in Packaging and Disposable Consumer Goods

HDPE	Containers for milk, dairy products, and ice cream; laundry detergent, bleach, and household cleansers; motor oil; paint; bottled water; basecups of PET soft-drink containers.
HDPE film	Grocery sacks and merchandise bags; bag and box liners of food products.
LDPE/LLDPE film	Wrapping for baked goods, candy, dairy, meat, poultry, seafood, produce; grocery sacks; dry cleaners' bags; shrink wrapping; trash bags.
PET	Soft-drink containers; dual-ovenable trays.
PS foam	Clamshell containers; hot-drink cups; disposable plates; egg cartons; meat and poultry trays; packaging beads.
Solid PS	Produce baskets; tumblers and cocktail glasses; disposable cutlery; disposable lids; dairy containers.
PVC film	Meat and poultry wrapping; shrink wrapping.
Rigid PVC	Containers for cooking oil and bottled water.
Composites	Squeezable condiment containers; aseptic juice cartons; bags for snack foods (potato chips); toothpaste tubes.

materials with high recycling histories, such as glass and paperboard. The reasons cited for these changes include increased convenience to the consumer, longer shelf life, and lighter weight. But the disposability of the packages and the environmental impacts of disposal have been largely overlooked. As complex combinations of plastic become more common in packaging, the potential recyclability of the solid-waste stream will decrease, since there is no recycling mechanism for these combination packages.

Building and Construction Materials

In 1985 the building and construction industries consumed about 10 billion pounds of plastics. It is expected that by the

year 2000 the industry will consume almost 15 billion pounds
of plastics in the form of pipes and fittings, plumbing and bath-
room fixtures, interior/exterior building materials, and air-sup-
ported structures.[10] Unlike packaging, these uses of plastics do
not immediately impact the waste stream, but their eventual
disposability must be considered in waste management planning
for the intermediate and long term.

The largest building/construction use of thermoplastics is
for pipe and conduit applications. Almost one-half of the polymers consumed in building and construction is used for piping.
PVC, HDPE, LD/LLDPE, and PP are used in potable water pipe,
irrigation pipe, drainage pipe, gas pipe, and various other constructions. PS is used for products such as light fixtures and
ornamental profiles. PVC is used for window profiles, flooring
gutters, foam moldings, and weather stripping, and is the major
plastic used in construction.[11] Uses of plastics in durable construction and consumer products are listed in table 4.

Postconsumer plastic waste from the building and construction industries totaled approximately 710 million pounds in
1984, or 2.4 percent of all postconsumer plastic wastes. About

TABLE 4
Plastic Resins in Durable Products

HDPE	Toys; housewares; piping.
LDPE	Piping.
PET	Magnetic recording film.
PP	Furniture; housewares; luggage; toys; automobile battery cases.
PS	Air conditioners and refrigerators; audio cassettes; radio, stereo, and television cabinets; furniture; toys; housewares; building and construction materials.
PVC	Flooring; paneling; automobile upholstery and trim; toys; footwear; luggage; credit cards; tablecloths; floppy disk jackets; upholstery; shower curtains; garden hose; piping; window profiles.

Source: "Materials '88," *Modern Plastics,* January 1988.

80 percent of these materials were thermoplastics, with PVC and LDPE the most common (46.5 percent and 10.1 percent, respectively); PS was present at 6 percent. Thermosets comprised approximately 20 percent of postconsumer plastic wastes, the most common being phenolics (13 percent) and polyester (6.5 percent).[12]

Postconsumer plastic wastes for the building/construction sector are projected to be 3.9 billion pounds, or 8.9 percent of all postconsumer plastic wastes, by 1995. The growth in the use of plastics for building/construction is projected to be large through the year 2000—surpassed only by the use of plastics in the packaging industry. The building/construction uses of plastics have a life span of 25 to 50 years, the longest of all categories of use. This means that plastic construction wastes present in the waste stream in 1984 were primarily from products manufactured in the 1950s. Thus the "revolution" in the use of plastics for building/construction that began in the 1960s will begin to impact the waste stream in the 1990s and in the next century.[13]

Consumer and Institutional Goods

A wide variety of consumer products use plastics: toys, housewares, disposable cutlery, sporting goods, footwear, medical and health care devices, laboratory supplies, lawn and garden instruments, serviceware, and luggage (see table 4). In 1984 about 4 billion pounds of plastics were used in consumer products; projections show this figure rising to 4.4 billion pounds in 1990 and 4.8 billion pounds in 1995.[14]

Consumer and institutional products are estimated to have a life span of five years. In 1984 this segment accounted for 12.3 percent of the postconsumer waste stream, or 3.6 billion pounds. Ninety-five percent of these materials were thermoplastics. PS, the most common material, constituted an estimated 20.6 percent of waste from this category, or 740 million pounds. Other plastics in the waste stream from consumer goods included LDPE (16.3 percent), HDPE (13.1 percent), PP (13.3 percent), and PVC (11.8 percent). Projections for 1995 estimate that consumer products will account for nearly 10 percent of postconsumer plastic wastes, or 4.4 billion pounds.[15]

Electrical/Electronic Products

The electrical/electronic market includes home and industrial appliances, electrical and industrial equipment, business machines, computers, records, tapes, and batteries. Resins commonly used in this market include polyethylene, PVC, PS, and PP. In 1984 some 2.8 million pounds of plastics were used for these applications; consumption is projected to increase to approximately 2.9 million pounds by 1995.[16]

Products in this area generally have a life span of about 15 years. Therefore, materials used in the manufacture of electrical and electronic equipment in the mid 1980s can be expected to appear in the waste stream around the year 2000. During 1984 this category of economic activity accounted for roughly 8 percent of all postconsumer plastic wastes, or 2.4 billion pounds. By 1995, this category of waste is projected to increase to 2.8 billion pounds.[17]

Furniture and Furnishings

The furniture and furnishings market consists of residential, office, commercial, and institutional furniture, as well as carpets, rugs, wall coverings, bedding, curtains, and blinds. The principal plastic materials in this market are polyurethane foams and PP, used in upholstery and carpets. Wood still dominates the furniture market, but there is an increasing trend toward the use of PVC laminates, particularly for unassembled furniture such as home entertainment centers, shelving, and computer stands.[18]

In 1984 furniture and furnishings manufacturers used roughly 2.2 billion pounds of plastics. Consumption is projected to increase to about 2.4 billion pounds in 1995.[19] These products have an approximate life span of ten years. Postconsumer plastic wastes from this category comprised 7.4 percent of the total postconsumer plastic waste stream, or 4.4 billion pounds, during 1984. Of this, 35 percent was PP and 28 percent polyurethane foam; other materials included PVC (7.3 percent) and urea and melamine (7.8 percent).

Projections for 1995 estimate that products from this category will comprise 7.2 percent of the postconsumer plastic waste stream, or 6.3 billion pounds. Polyurethane foam and PP are expected to continue their dominance.[20]

Transportation

The transportation market includes automotive, marine, railroad, recreational, and military applications. The use of plastics in the transportation sector is expected to decrease slightly from its 1984 level of 2.1 billion pounds to 2 billion pounds by 1995. Although plastics are anticipated to constitute a larger portion of the typical automobile's weight in the future, recent trends in the use of plastics in automobiles and the production index for transportation equipment indicate that the use of plastics in overall domestic transportation equipment will not increase in the coming decade.[21]

In 1984 postconsumer plastic from the transportation sector totaled roughly 1.9 billion pounds, or 6.7 percent of the nation's postconsumer plastic waste. Fifty-two percent of these waste materials were thermoplastics, including PP (17 percent) and

PVC (12 percent); 30 percent were thermosets, with polyester the most common (24 percent); and polyurethane foam accounted for the remaining 18 percent. Projections for 1995 estimate plastic wastes from the transportation sector to be 5.2 percent of postconsumer wastes, or 2.2 million pounds.[22]

Plastics use continues to increase in automobile manufacturing, where new grades of plastics are being introduced for use in body panels, bumpers, and interior components. Thermoplastics are becoming increasingly common in automobile design.[23] The use of plastics for bumper systems in North American–made vehicles is expected to increase by at least 60 percent in the next decade. Nearly 90 percent of the cars sold in the U.S. in 1990 are expected to have one or more major plastic components in their bumpers.[24] By 1992 the typical U.S. car is likely to contain at least 400 pounds of plastics and composites, up from 200 pounds in 1984.[25]

Industrial Machinery

The industrial machinery market includes signs and displays, farm and construction equipment, machine tools, marine supplies, and engine parts. In 1984 this sector of the economy used 0.4 billion pounds of plastics; usage is expected to decrease to 0.3 billion pounds in 1995.[26] In 1984 postconsumer plastic wastes from this category comprised 1.2 percent of total postconsumer plastic wastes, or 360 million pounds. HDPE, PP, nylon, and phenolics were the predominant materials. In 1995 this sector is estimated to comprise 1 percent of postconsumer plastic wastes, or 433 million pounds.[27]

Adhesives

Adhesives are used as alternatives to mechanical fasteners in automotive, aerospace, and other structural applications; for bonding dissimilar materials in multilayer constructions; and for lamination and coextrusion of polymeric building materials and weatherable thermoplastic panels.[28]

Use of plastics in the manufacture of adhesives reached 2.5 billion pounds in 1985, with the majority being vinyl (34 percent), thermosets (30 percent), and styrenics (23 percent). Usage is expected to grow to nearly 3.6 billion pounds by 2000, with thermosets comprising a 32 percent share, vinyl 29 percent, and styrenics 24 percent.[29]

2

Focus on Packaging

Although the use of plastics is increasing in almost all sectors of the U.S. economy, the most rapid growth is occurring in the manufacture of packaging materials.[1] The packaging portion of the municipal waste stream has become the focus of attention for solid-waste planners, legislators, and environmentalists because it is from packaging that major gains in recycling must occur if the nation is to have an effective integrated waste-management system.

As plastics proliferate in the packaging sector, they inevitably take the place of glass, metal, paper, and paperboard packaging. Because well-secured recycling systems already exist for most of those traditional materials, but not for plastics, there is increasing concern over the current recyclability of the waste stream and the nation's ability to meet local and state targets for recycling that are in line with the current preferred hierarchy of waste-disposal options. This hierarchy—waste reduction, recycling, resource-recovery incineration, and landfilling, in that order—has been incorporated into New York State law and has been accepted by the U.S. Environmental Protection Agency in its solid-waste planning. The preference for waste reduction and recycling over more controversial waste-disposal options such as incineration and landfilling has led to increased expectations for recycling. But if, at the same time, the recyclability of the

27

packaging waste stream diminishes, it may become impossible to meet targets currently being set in statute.

As noted in chapter 1, plastic packaging waste constitutes approximately 4 percent by weight of the entire waste stream.[2] The impact of plastic on the waste stream, however, cannot be judged primarily by weight because plastics are so light in comparison to their volume. In landfills, for example, plastics occupy much more space relative to their weight than do most other materials.

The combined segments of the packaging industry make up one of the largest manufacturing groups in the United States. Total shipments of packaging are about 4 percent of the value of all finished goods in the U.S. Shipments of the industry in 1985 were $55.8 billion, up from $43.5 billion in 1979.[3]

In 1980 approximately 14 percent of all discarded packaging was recycled. Most of this packaging was paper and paperboard, with a recycling rate of 25 percent, and aluminum was recycled at a rate of approximately 29 percent. The plastic recycling rate was nearly zero.[4]

A study conducted for the United States Department of the Interior (DOI) in 1987 forecast the substitution of new plastic materials for conventional metals and glass in major U.S. industrial sectors in the 1990s. In the U.S. packaging industry, the DOI study found that plastics were displacing significant amounts of both glass and metals, as well as competing with paper and paperboard.[5]

The authors identified the soft-drink container market, which accounts for about one-third of all can and bottle demand, as one of the major areas of substitution. The penetration of plastics into this market is primarily due to their lighter weight and low susceptibility to breakage, attributes that reduce handling costs and fuel consumption costs in shipping. In order to compete, glass-bottle manufacturers began using thinner glass-walled soft-drink bottles covered with plastic foam in 16-ounce and one-half liter sizes.

The DOI study estimated that 75 percent of all metal cans

used by the packaging industry are for beverages and 25 percent are for food. Aluminum has 94 percent of the beverage can market, and steel the remaining 6 percent. The reverse is true for food cans, where steel has 94 percent of the market. However, the food can market is expected to decline steadily, partially due to the increased use of frozen-food packaging in paper and plastics and the increased popularity of microwave ovens, which has spurred the development of multimaterial (and largely unrecyclable) microwaveable packaging materials.

In the rigid-container sector, the DOI study forecast that by 1990 plastics will displace approximately $386 million in market share held by glass containers and $223 million by metal containers. This displacement accounts for about 7 percent of the glass market and 2 percent of the metal container market (based on 1985 data).

Overall use of plastic bottles rose from 75 million pounds in 1960 to over 2 billion pounds in 1984. In 1985 over 18 billion plastic bottles were produced in the U.S. Seventy-two percent of plastic bottles are used as food and beverage containers.[6]

The art of package design has undergone dramatic changes in the past few years. New technologies and material combinations allow plastics to be used in ever more applications. For example, the development of barrier packaging technology has greatly expanded the number of foods that can be packaged in plastic. Food manufacturers and package manufacturers work together to custom-design packages for specific applications. Plastic is often the material of choice because it is lightweight, microwaveable, durable, and resistant to breakage.

One of the newest trends in packaging is composite containers, receptacles made of several different materials. The development of high-barrier polymers and multiple-polymer containers has enabled plastics to be used in applications requiring oxygen, carbon dioxide, flavor, odor, and solvent protection. These multilayer composite containers usually consist of structural layers bound to a barrier layer by adhesives. Resins used in the structural layers include PP, polycarbonate, and PET.

About 300 million units of high-barrier multilayer coextruded containers were produced in 1985. By 1995 production is estimated to soar to 29 billion units a year, an average annual growth rate of 58 percent.[7]

Other trends in food packaging include microwaveable/dual-ovenable packages, aseptic packaging, and squeeze bottles. Aseptic packages, made of aluminum foil laminated with plastic, are capable of keeping beverages such as milk and fruit juice drinks fresh for six months without refrigeration. These containers are becoming increasingly popular with consumers.

Flexible packaging materials include plastic bags, liners, aluminum foil, paper, and combinations. Plastic is rapidly replacing paper in the grocery-sack market and foil in the dual-ovenable market. More often, plastic is being combined with foil and paper for packaging of snack foods such as potato chips and cookies.

Other recent changes in the packaging of food and beverages include:

- The complete phase-out of glass milk bottles in favor of polyethylene-coated paperboard and HDPE bottles.
- The substitution of "shelf-stable," microwaveable, "table-ready" multilayer plastic containers for metal cans and glass containers.
- The use of PET containers for peanut butter and mustard, replacing glass.
- The manufacture of cereal box interior liners made of PP instead of paper or cellophane.
- The use of dual-ovenable plastic containers, instead of aluminum, for frozen foods.
- The replacement of all-paperboard containers by a combination of paperboard and LDPE.

One setback for the plastics manufacturers was public resistance to the "plastic can." This container, manufactured by the Petainer Corporation, is made of a PET body, aluminum top, and PVC shrink-wrap exterior. It was introduced by Coca-

Cola Company in a Georgia test market in 1986 and by the Original New York Seltzer Company in test markets in Michigan, Massachusetts, and New York in 1988. In each case, the "plastic can" was met with a storm of opposition from recyclers and environmentalists because of its potential to replace the highly recyclable aluminum can. A highly visible symbol of current trends, the "plastic can" became the catalyst for general opposition to new packaging design overall. Opponents were successful in convincing both companies to abandon, at least temporarily, the use of this container. (Further details of this controversy are discussed later in this chapter.)

Package designers have not been required to consider disposability and environmental impacts of discarded packages when designing new containers. Generally, their areas of expertise are engineering, design, and marketing, not waste disposal or environmental protection. Therefore, many "exciting" new packages inevitably become disposal problems once their useful lives are complete, being neither recyclable nor naturally biodegradable.

Some designers, however, are gradually becoming aware of their role in the problems of waste disposal. Stuart Mosberg, a well-known packaging designer in New York City, described his new understanding at a conference sponsored by the New York State Legislative Commission on Solid Waste and by the Northeast Governors Conference in 1989. His realization came during the now-famous summer of 1988, when the memory of the garbage barge *Mobro* was fresh and the problems on nearby beaches were acute. Mosberg's decision to take a leadership role in his industry may mark a turning point for designers, as well as for the design students now in schools around the country.

Why the Change?

With the advent of the "disposable society," the waste stream has undergone dramatic changes. Many reasons have been given for the shift to the use of plastics: "changing lifestyles," the desire for convenience, a trend toward smaller families, an aging population, and rising employment. These social changes coupled with technological developments in package design have spurred production of consumer products in single-service packages, many made of plastic for microwave ovens. Indeed, probably no other technological development has created more changes in food packaging than the microwave oven, which requires nonmetal containers. Its popularity has led to a major proliferation of plastic packages designed to go from freezer to microwave to table, and then to the garbage pail. Nearly 70 percent of the homes in the United States, it is estimated, now have a microwave oven.[8]

At the same time, growing concerns over product tampering have led to the increased use of protective layers of packaging. And the disposal of plastic and paper trays, utensils, and containers is viewed by many consumers and institutions as quicker, easier, and sometimes more sanitary than cleaning, storing, and reusing serviceware.

For the manufacturers, the lighter weight of plastic reduces freight and energy costs. Manufacturers also cite reduced losses due to less breakage on the production line. Convenience advantages to the consumer also include lighter weight and shatter resistance. For package designers, plastic offers exciting alternatives because it can be formed into a variety of shapes and, as mentioned earlier, the new barrier technologies allow plastics to be used for a growing array of foods.[9]

Package Design

Designing a new package is a complex and lengthy process; developing a new package may take two to three years. During that time the producer of the product works with the package designers. The process usually begins with a market research study to identify demographic trends and target markets. Issues investigated include family size and types (e.g., single parent, dual-career vs. single-career households), buying patterns (e.g., large weekly shopping trips vs. more frequent shopping trips for a few items), eating habits, and microwave-oven ownership. Based on the results of the research, market niches are identified for the development of new products of all kinds. Marketing departments decide how they want the product to be contained and displayed; specifications are then drawn up for the desired new package.

Major issues addressed in the design process include:

- How will the product be marketed?
- What will the package cost?
- What level of protection is desired?
- What shelf life is desired?
- How will the packaging material affect the product?
- For food products: Is the product to be cooked in the package?
- Can the package be manufactured using existing equipment?

Although the most important function of packaging is to protect and preserve the contents during transportation and storage, packaging is also an advertising medium: the visual appeal of the package is intended to influence the consumer's perception of the product. Food processors and other packagers are increasingly relying on packaging design as a source of product identification and as a means for gaining a competitive edge over similar products.

Current package design research involves a number of testing methods. Nine research methods are now widely used:[10]

1. A focus group provides manufacturers with insights into consumers' opinions, attitudes, and preferences. The group, a small gathering of consumers under the guidance of a moderator, engages in a discussion of a selected topic.

2. The personal interview asks consumers to respond to a questionnaire.

3. A Tachistoscope (T-scope) test measures the visual impact of a package, which is shown on a screen for a fraction of a second. Respondents are asked to describe what they saw. The package is shown repeatedly, increasingly for longer exposures, until respondents identify all salient elements.

4. An eye-tracking test indicates how effectively the design elements of a package are combined and arranged. A camera mechanism is trained on the respondent's eyes as he or she views the test package. The camera records the sequence in which the various design elements are viewed and how much time is spent on each element.

5. The angle meter simulates the view of a package as seen by a consumer walking down a store aisle. The respondent is asked questions about a package shown at an angle rather than front-facing.

6. The blur meter registers an out-of-focus slide, in simulation of the fact that many shoppers with corrected vision do not wear their glasses while shopping, and respondents are asked to identify a package.

7. The semantic differential test is used to explore how the package affects product perception. The respondent is shown a package and asked to evaluate the product in terms of preselected traits (e.g., freshness, wholesomeness, quality, value) and on a predetermined scale, either numerical or verbal.

8. Forced-choice tests ask respondents to select one of two products (usually the same product in different package designs). The premise is that the determining factor is the packaging, since the respondents have no other basis for choosing.

9. The miniature store test simulates how a design performs under purchase conditions. Respondents are given an amount of money and told to spend it as they wish inside a specially erected store. Among the offerings is the product category containing the test package design.

From market research of this type, for example, Mel Druin, vice-president of packaging for Campbell's Food Company, concluded that microwaveability and table-readiness are desired attributes in a container. Before switching from aluminum to plastic in the frozen dinner market, Campbell's conducted consumer research and found a 7-to-1 preference for plastic.[11]

However, Dr. Dan Toner, senior scientist for Campbell's, admits that the company is concerned about the present and future choices guiding company policy and concedes that the nonrecyclability of much of the new packaging is a factor in decision making. The PET microwave tray, he notes, is recyclable. The multimaterial nature of the package as a whole, however, makes true economic recyclability problematic.[12]

An even newer development is the microwaveable shelf-stable entree line, a category of food that is becoming increasingly popular. These entrees can be displayed on the grocery shelf, rather than in the frozen foods section, which tends to be less visible to shoppers. "Shelf-stable" foods can be microwaved and eaten in the same container, which is then immediately disposed of. Hormel's Top Shelf, one of the first unfrozen, "shelf-stable" microwaveable entrees, is an example of the opening up of this new market. The entrees are vacuum-packed in a multilayered plastic serving dish and sealed airtight with a peel-off lid; shelf life is said to be at least 18 months.[13] The product is targeted at consumers between the ages of 25 and 54 with active lifestyles and annual incomes of more than $30,000. Top Shelf entrees were developed to fill Hormel's niche in preprepared "homemade-type" foods. Once Hormel identified this niche, it entered into a joint venture with a package supplier to develop a suitable packaging system.[14]

Often a package is designed with a specific food product in

mind, but in some cases, the reverse is true. When American National Can Company developed a coextruded squeezable bottle, the company approached Heinz with the suggestion that the new bottle could be used for packaging ketchup. The selling points included "squeezability," reduced shipping costs through lighter weight, and less breakage on filling and handling lines.[15] Heinz has now dropped multi-layer, multi-resin squeezable bottles in preference to PET mono-resin squeezable bottles.

Some companies will not discuss how they make decisions on packaging. For example, Best Foods, the maker of many familiar brands of products, including Skippy peanut butter, will not reveal if environmental considerations play a part in their process. After "consumer demand" and "consumer testing," Best made a decision in early 1988 to change most of the Skippy containers from glass jars with metal tops to plastic jars with plastic tops. In a letter of January 23, 1989, to a customer, the company Supervisor of Consumer Response, Eileen E. Mason, reported that testing revealed that "plastic packaging was overwhelmingly preferred over glass for peanut butter — a product used extensively by children." Other Best Foods' representatives call the new package desirable because it is made of PET, which can be recycled. But the company feels no responsibility to create or support a recycling infrastructure, believing that to be the responsibility of local governments.[16] When asked, company officials do admit that they have switched from a highly recyclable and reusable glass container to one of uncertain recyclability, at least at present.[17] However, representatives of the company seem open to discussion and say they are closely monitoring consumer response to the plastic Skippy jars and other packages.

High-Visibility Consumer Plastics

Certain high-visibility plastic products have become emblematic of the proliferation of plastics throughout the consumer economy. One of these is the ubiquitous grocery sack, long manufactured from paper, increasingly made of plastic.

The plastic grocery-sack market is estimated at 22 billion pounds a year, or about 400 million pounds of polyethylene. Plastic sacks have moved from virtual nonexistence in 1980, to almost 5 percent of the market in 1982, to one-quarter in 1985. Annual market growth has been estimated at 25 percent.[18]

Plastic grocery sacks are composed of polyethylene, either LLDPE, which has 60 percent of the share, or high-molecular-weight HDPE, which has 40 percent. Additives used during manufacture include calcium carbonate, an inert filler; talc, an antiblocking agent (to prevent the adjacent layers of plastic from adhering to each other under pressure); slip agents (lubricants that aid in the manufacture of plastic); an antioxidizing agent; and a colorant. All additives, as well as the inks used in plastic grocery bags, must meet U.S. Food and Drug Administration (FDA) criteria for food-contact chemicals. Thus the inks used on grocery bags are water-based. However, nonfood plastic bags need not meet these standards. Colorants and inks on nonfood bags may include chrome yellow, cadmium yellow, cadmium orange, mercury cadmium (red), and cobalt blue.[19]

The decision by a supermarket chain to switch from paper to plastic bags is usually based on plastic's lower cost per unit. Consumer response to plastic bags has generally been mixed. Many consumers prefer paper bags because they retain their shape better when placed in an automobile. Also, plastic bags do not hold as much as paper bags, requiring the use of more bags per customer. However, some consumers, especially those in urban areas who carry their bags, prefer plastic bags because they have handles. Increasingly, stores in urban areas have created a combination bag—a paper bag in an outer plastic bag—but this increases the cost to retailers.[20]

The first study comparing paper and plastic grocery bags was conducted by the Midwest Research Institute for Mobil Chemical Company in 1980.[21] This analysis compared resource and environmental impacts of polyethylene and Kraft paper grocery sacks, beginning at the point of raw material extraction (crude oil production, wood harvest, and natural gas production), and including material processing, manufacturing, and disposal of sacks. Seven impact categories were analyzed: raw materials requirements, energy input requirements for manufacture, waste-water volume from manufacturing, industrial solid wastes, atmospheric emissions during manufacture, waterborne wastes from manufacture, and postconsumer solid wastes. The study concluded that, overall, plastic sacks showed fewer impacts in these categories when compared to paper grocery bags. The greater impacts of the paper bags were attributed to the pulp and paper manufacturing steps. Energy input requirements were similar for both bags. For paper, most is in the form of process energy required for pulp processing and paper production; for plastic, more than half is material resource energy (hydrocarbon feedstocks).

In the postconsumer solid-wastes category, plastic was found to require less landfill space: 1 cubic foot of landfill can accommodate 60 pounds of plastic, but only 44 pounds of paper.

The study, however, did not address at least four key points:

- How would these impacts change if recycled fibers were used to manufacture both types of bags? (The use of secondary materials, both for paper bags and plastic, saves energy and natural resources.)
- What are the long-term environmental impacts of landfilling?
- What are the environmental impacts of incineration, composting, and recycling for each type of bag?
- What are the environmental impacts of improperly disposed bags of each type?

In 1984 the American Paper Institute commissioned a study to compare the handling costs of paper and plastic grocery bags.[22] The study concluded that using paper bags can be considerably less expensive for supermarkets than plastic bags. In certain situations, the study found, the savings of paper bags over plastic bags can range as high as 10 percent, and the average saving was found to be 3–6 percent or about 15–30 percent of the cost of the bags. Front-end labor (checking and bagging) was by far the largest cost associated with either type of bag, accounting for 80 percent of the total cost.

The larger numbers of items sold per retail customer, the study found, the greater the advantage of paper bags, due to their larger capacity. The advantage shifted even more heavily to paper when checker/bagger teams are used, because paper provides a better opportunity to use baggers productively. Other factors that affected the comparative costs included price per bag, types and sizes of bags available at the checkout counter, policy for deploying baggers, checkstand design, extent of double bagging, and wage rates. However, the study found that the total costs of using paper bags were always less than those of using plastic.

It should be noted that both of the above studies were undertaken for interested parties: Mobil Chemical Company and the American Paper Institute. Therefore a study, discussed later in this chapter, undertaken by the Shop-Rite supermarket chain should be deemed to be more independent.

Another study—again commissioned by an interested party—was performed by the Midwest Research Institute for Mobil Chemical Company to examine and compare the total environmental impact of meat trays made of PS foam and those of molded wood pulp.[23] This study, conducted in 1972, looked at comparative environmental impacts, starting from raw materials extraction through production, delivery, consumer use, and final disposal.

Production of both types of trays requires energy and results in pollutant emissions. Atmospheric emissions may include particulates, sulfur dioxide, nitrogen dioxide, carbon monoxide, and hydrocarbons. For both types of trays, process waste water contains suspended solids and biological oxygen demand as well as carbon oxygen demand and oil for polystyrene trays. Process solid waste includes sludge, ashes, and scrap. Input energy includes coal, oil, gas, and electricity for manufacturing processes, as well as raw materials from hydrocarbons for PS.

The study concluded that PS trays resulted in lower impacts in all categories: raw materials consumed, energy input requirements, waste-water volume from processing, solid waste, and atmospheric emissions and waterborne wastes from production.

Postconsumer solid waste was considered separately in the study. Foam trays are inert in landfills, while pulp trays are biodegradable. No advantage was assigned to either, and neither was deemed to be a particular problem in land disposal. Pulp trays, however, were found to occupy approximately 20 percent less volume than foam trays. Neither energy considerations for transportation of waste to disposal facilities nor tipping fees of waste were considered.

Degradability was considered a positive factor in composting, but since composting applied to less than 1 percent of the waste stream at the time of the study, degradability was not considered significant. Although composting is still not a popular method of waste management, it is now being considered much more seriously as a viable strategy; therefore, this finding may have greater significance today than at the time the study was completed. Recycling was not addressed, since neither tray was being recycled at the time of the study. Today, in contrast, both pulp and PS pilot recycling projects are being undertaken.

As mentioned earlier, another highly visible — and controversial — symbol of trends in package design was the trial introduction of the so-called plastic can, manufactured by the Petainer Company. Recyclers and environmentalists saw the

plastic can as aimed directly at the most highly recycled component of the packaging stream—the all-aluminum can. In response to the negative publicity and organized opposition, Coca-
Cola discontinued its test marketing, citing the lack of a
recycling infrastructure for the can. Nonetheless, in 1988, the
Original New York Seltzer Company decided to test-market the
plastic can in the nine "bottle bill" states. Introducing the can
into markets where deposits and returns would collect masses
of the cans for possible recycling was judged to be a safer trial
than the one used by Coca-Cola. But public outcry was even
louder, and bills were introduced in some states banning this
particular package.

So, Original New York Seltzer also abandoned the plastic
can, despite what it described as favorable consumer reaction
and a pledge to recycle the can through the PET recycling infrastructure. But the expense involved in recycling a package
made of three different materials made recycling this particular
package economically unfeasible. Negative publicity was also
a strong deterrent to further marketing, the company acknowl-
edged, as were letters from teachers and students who had been
alerted to the issue by such organizations as Environmental
Action Coalition, Environmental Action, and the Coalition for
a Recyclable Waste Stream (a national consortium of environmental and recycling groups).[24]

In New York State, it was estimated that the introduction
and use of the plastic can in place of the all-aluminum can for
beverages might have seriously disrupted the overall recycling
rate, since 82 percent of aluminum beverage containers, but
only 57 percent of plastic beverage containers, are recycled in
that state.[25]

A final example of a highly publicized controversy has been
McDonald's decision to continue using PS products in the face
of consumer pressures and some legislated bans. Highly criti-
cized for its volume of throwaway packaging, McDonald's has
become the main target of those who would prefer to see fast-
food packaging rely on paper. McDonald's well-known "clam-

shell" container and the company's reliance on PS cups led McDonald's to attempt its own collection and recycling campaign. Its 1989 plan focused on its previously announced pilot recycling effort in Greenpoint, Brooklyn, a facility that accepted the mixed trash from 20 McDonald's outlets in New York City and Long Island, sorted and rinsed the PS, and sent it off by truck to an Amoco Foam pelletizing plant in New Jersey. Amoco Foam sent the PS pellets to its manufacturing plant in Winchester, Virginia, where the recycled PS pellets were mixed into batches of virgin pellets for use in such products as insulation. The recycled PS pellets were supposedly "preferred by the Winchester plant because of a 20¢–30¢ advantage per pound over virgin resin," according to a company representative.[26]

Due to increasing public pressure, in late 1989 McDonald's helped to create an industry recycling consortium for the post-consumer collection, processing, and recycling of PS foam from their outlets. At "dialogue sessions" with environmentalists, including those from EAC, the company promised that all 450 of its New England outlets would begin creating what would become a nationwide infrastructure. This process was underway by early 1990.

However, crucial issues are still unresolved, including worker safety in the production of PS foam and the creation of a community recycling infrastructure for materials not used and collected in the restaurants themselves. By April 1990, in a move indicative of the continued pressure, the New York City Council held hearings on a proposed bill to ban the use of PS foam in almost all retail applications.

Some Changes in Packaging

As more public attention focuses on packaging, and more packaging legislation is enacted in municipal and state legislative bodies around the nation, the affected industries are beginning to assume more responsibility for their decisions.

Until very recently, disposal aspects have not been part of the criteria for decisions by either the product maker, the manufacturer, or the package designer. Now, many companies are in the process of reevaluating their strategies regarding disposability and recyclability. An example of such an evaluation is the decision to develop an all-PET beverage container—one that does away with the basecup made of HDPE. Although HDPE is recyclable, the need to separate the HDPE from the PET makes the recycling of combination containers more expensive. Also, HDPE is currently recycled at a much lower rate than PET.[27] Another example of environmentally responsible change is the decision of the Yo-Plait Company to switch back to paper containers from plastics on the grounds that paper cups are "more biodegradable."[28]

A potentially very important change is the decision by many plastics producers to put codes on their packages. As early as 1984, the issue of imprinting codes on plastic items was presented to the Society of the Plastics Industry (SPI), a trade organization. At that time, SPI fiercely resisted the idea, calling it "unfeasible."[29] The absence of codes, however, made it virtually impossible to recycle plastic packaging, even when the will to separate and the ability to market small amounts of plastics were already in place.

Currently, with the exception of PET beverage containers—where "common knowledge" of the package's resins has led to collection—any separation by voluntary recycling centers, organized curbside pickup, or apartment buildings relies on guesswork. In the Environmental Action Coalition's laundry detergent and bottled water container recycling project in apartment buildings, the collectors assume that these containers are made of HDPE, as "commonly known."[30]

As SPI and its newly formed association, the Council for Solid Waste Solutions, begin to promote the voluntary coding system, it is expected that many more plastic items will be coded. This, in itself, will foster the long-needed infrastructure for economically viable separation, collection, and recycling of certain plastics. (For more information about SPI's coding system, see chapter 4.)

Although most supermarket chains are waiting for the manufacturing and production segments of the economy to effect changes, the Shop-Rite Supermarket Company has begun its own initiatives. All stores in the chain have been recycling their corrugated cardboard distribution packaging for years, and more recently they began recycling plastic shrink wrap from shipping pallets. As of early 1988, the chain had recycled more than 77,000 pounds of plastic packaging. In its private-label products, the chain uses recycled paperboard "wherever possible."[31]

Shop-Rite is also investigating alternatives to foam packaging for meat and has stopped using plastic wrap in its bakery operations, substituting waxed paper. According to Mary Ellen Gowin, Vice-President for Consumer and Public Affairs, Shop-Rite believes that if retail stores take the lead, they may be able to pressure food companies to rethink their packaging decisions.

Environmental Action Coalition (EAC) Surveys

The unavailability of waste-stream packaging data from the New York City Department of Sanitation and the inability to verify figures taken solely from industry sources led EAC to conduct its own primary data survey in selected supermarkets and drugstores in New York City. The surveys took place in spring 1988, under the supervision of Science Research Associate Ellen Feldman. The first survey was conducted for EAC by the New York City Volunteer Corps (CVC), a group of young people, 17 to 20 years of age, serving in a full-time one-year program sponsored by the City of New York. (CVC organizes its volunteer force to provide a variety of services for city agencies and nonprofit organizations, thus providing its volunteers with specific training that will lead to gainful employment. Many of the young volunteers go on to higher education, also sponsored by the CVC program.)

The survey form listed approximately 75 products commonly found in supermarkets. The surveyors were to indicate on the form the packaging material(s) used in products they saw on the supermarket shelves. Five supermarket chains were surveyed throughout Manhattan, Brooklyn, the Bronx, and Queens: C-Town, Key Food, Waldbaum, Met Food, and Associated. In all, 63 stores were surveyed; permission was granted either by the chain's main office or by store managers. Despite EAC's efforts to expand the number of chains surveyed, many of the larger chains refused to participate, describing the issue of food packaging as too controversial.

Major findings of the EAC supermarket survey were:

- Over 50 percent of all the items surveyed were packaged entirely in plastic, either single layer or multiple layered.
- Food categories in which nearly all items were packaged in plastic included dairy, produce, meat, and household cleaning items.
- Approximately 6 percent of all items were packaged in a combination of materials that included some plastic.
- Approximately 33 percent of all items surveyed were packaged in glass; 11 percent were in paper or paperboard.

EAC's drugstore survey was conducted by the high school and college students in the Training Student Organizers (TSO) Program of the Council on the Environment of New York City, a not-for-profit group located in the Office of the Mayor. Forty-one drugstores were surveyed, including small neighborhood stores as well as chain stores such as Genovese, Pathmark, Duane Reade, Walgreen, Rite Aid, and CVS. Most of the stores were located in the Bronx and Brooklyn.

Major findings of the EAC drugstore survey were:

FIGURE 4
Survey of Plastic Packaging in Drugstores

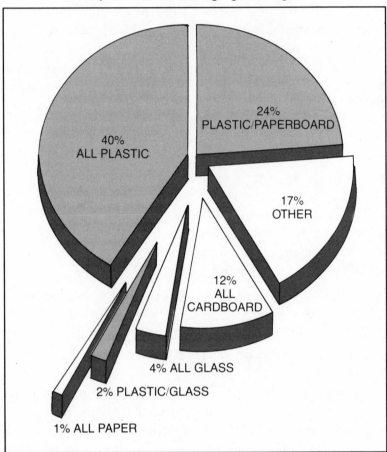

Source: Environmental Action Coalition and Council on the Environment of New York City Surveys, 1988.

- Over 75 percent of products surveyed had all or some plastic packaging.
- Nearly 40 percent of products had all-plastic packaging.
- Only 1.3 percent of the packaging was all paper; 3.9 percent was glass or some glass; 12.1 percent was all card-

board; 24 percent was plastic and cardboard combination, with most cardboard boxes containing plastic bottles or tubes.

Implications of these two surveys bore out EAC's assumption that the packaging waste stream—from which large percentages of recyclables must come to meet any city or state recycling goals—has changed from metals and glass of long-sustained and easy recyclability to products having little, if any, recyclability. Information obtained from contacts now participating in New York City's curbside program—which targets metals, glass, and newspapers—bears out the survey findings. In interviews, participants in Brooklyn Community Board 2 and Staten Island Community Board 3 expressed surprise and some dismay at the small number of recyclables available from households to deposit in the special containers provided by the Department of Sanitation. The households surveyed were buying the same basic products they always had, but the products are now packaged in plastic or multimaterial packages, not in glass or metal.[32]

Overall, EAC's surveys indicated that nearly 60–70 percent of consumer goods sold in supermarkets and drugstores are now packaged in all-plastic or part-plastic containers.

National studies estimate that plastic accounts for approximately 12–15 percent of the packaging waste stream by weight. The EAC survey did not analyze by weight, but rather by package units. Inasmuch as almost all of the packaging surveyed by EAC's teams was almost immediately bound for the waste stream, the change to plastic packaging does not bode well for estimated percentages of materials available to be separated as recyclables.

In order for communities to achieve mandated recycling goals, the target materials must be present in the waste stream. If they are replaced by nonrecyclable materials or materials not included in recycling plans (very few recycling plans at present include many of the types of packaging reported in EAC's surveys), the locally mandated goals for recycling will be impossible to meet.

Suggested Criteria for Packaging

Clearly, packaging designers need to consider recyclability, disposability, and solid-waste impacts in their work. While the following are not the only criteria to be considered, they should be among the most important:

- Packaging should be reusable. Reusable packages, such as refillable glass bottles that may never find their way into the waste stream, are ideal.
- Excess packaging should be eliminated. Excess packaging can be defined as packaging that exceeds the functional requirements of containing and protecting a product. For food products such requirements include the need to protect hygienic integrity, freshness, odor, and flavor.
- Packaging that creates harmful disposal should be eliminated. Certain packaging materials — not only plastics — have been shown to create health hazards to humans and ecological harm to the ecosystem through emissions from incineration or leachate from improperly managed landfills.
- Packaging should be made of a single material. Experience has shown that single-material packaging is much easier to recycle.
- Packaging should use recycled materials. Whenever possible, packaging — which almost always goes directly into the waste stream — should itself be made of recycled materials. Consideration should be given to banning the use of materials for food packaging that physically and technologically cannot be reprocessed into food packaging. At present, FDA regulations for sterilization prohibit many materials from being reused as packages for food.

- Packaging should be fully and economically recyclable. Tiny recycling pilots that show recyclability of packages due to extreme measures do not prove true, economically viable recyclability.
- Packages should be clearly labeled as to contents of packaging materials and their recyclability. At present, consumers have no way of knowing what materials are used in packaging, and no legal standards define the packaging's true, economically viable recyclability.

In the absence of national commitment to the environmental examination of packaging, numerous local and state regulations have been passed or are pending. Clearly, in a nationally based economy, the examination and regulation of packaging for environmental control should come from the federal level.

3

Disposal of
Waste Plastics

All types of plastics present potential disposal problems in the solid-waste stream due to their inert properties, the use of additives and plasticizers, and the lack of widespread recyclability. The rapid proliferation and short life span of packaging, as discussed in chapter 2, has focused public attention on this segment of the postconsumer waste stream. In particular, disposables, such as six-pack beverage container rings, tampon applicators, fast-food service goods, and diapers, are increasingly seen as environmental problems.

The especially harmful presence of waste plastics in waterways led to federal action through the Marine Plastic Pollution Research and Control Act of 1987. As the provisions of this act come into effect, it will affect all aspects of plastics in waterways, including a provision forbidding the disposal of plastic waste from ships. Of special concern in this regard has been synthetic fishing gear (gill nets, set nets, and trawling nets), which is often discarded or lost at sea, where it lasts virtually forever.

Plastic debris from boats or from beach litter poses a serious threat to nearby marine animals, who either become entangled in the debris or ingest it. Sections of nets and six-pack rings have been particularly harmful to sea and shore birds, which can strangle themselves on these items. Ingestion of plastic balloons or balloonlike packaging has killed marine mammals as well. The decline in population of the northern fur seal (4–8 percent annually) is primarily attributed to death by entanglement.[1] Plastic materials, most often polyolefin sheeting and packaging film, have been observed in the stomach contents of whales and dolphins, and there is evidence to suggest that the death of some sea turtles is due to the ingestion of plastic bags, which the turtles mistake for jellyfish. There is also some evidence indicating that marine birds ingest plastics, sometimes resulting in stunted growth or death.[2]

While waterfront litter and illegal boat dumping present many hazards, many more potential dangers lurk in routine solid-waste disposal practices. The increase in the percentage of waste composed of plastics poses problems for landfills and for incineration plants. Data on these effects are scarce, and concerns are growing.

Landfilling

In landfilling, or dumping—still the most widely used method of waste disposal in the United States—waste is compacted, covered daily with a layer of clean soil fill (if properly managed), and capped with a thicker cover of earth when completed. Large fractions of waste in landfills gradually decompose and are converted to lower-molecular-weight solids, liquids, and gases. It is estimated that a mixed landfill needs to settle for 25 years before further use, most frequently as a park or golf course.[3]

Over time, ferrous metals in landfills are oxidized, and organic and inorganic wastes are broken down by microorganisms through aerobic and anaerobic processes. Food wastes tend to degrade most rapidly. Other materials, such as rubber and some demolition wastes, are quite resistant to decomposition. Glass, of course, does not degrade and is merely crushed into small particles. Plastics, as inert synthetic compounds, do not degrade.

Recent studies of landfills in Arizona have shown, however, that the compacting of wastes in landfills into "cells" with earth cover (allowing minimum oxygen and moisture to reach the waste) delays for many years the breakdown of even organic materials. Therefore, claims about the degradability of materials in landfills need to be examined in light of the management of each particular landfill.[4]

In a well-managed landfill, a number of factors promote a more successful life span and ultimate positive use after closing:

- Landfills managed as part of an overall integrated waste-disposal system should receive only materials that will break down; recyclables and nondegradables should be removed.
- Such landfills are managed for maximum decomposition, creating potentially valuable methane to be recovered as fuel and prolonging the life span of the landfill site.
- All wastes are covered daily to minimize odors and vermin.
- Maximum care is taken to control toxic leachate, which results from water percolating through the fill, often picking up contaminants that can pollute nearby waterways or underground water sources.
- At the end of the landfill's life span, earth cover is provided and settlement is allowed to occur.
- The best end use of a well-managed closed landfill is open space, since further settlement may take place on an irregular basis.

Given these criteria, opinions differ concerning the landfilling of plastic wastes. Because they are resistant to physical and biological degradation, or decomposition, plastics in a landfill remain inert indefinitely. If plastic bottles remain capped, they may balloon and rise to the surface, impeding the compacting of wastes. Wind-blown debris is also a problem, littering the nearby landscape and waterways.

Some landfill engineers and planners, however, believe that plastics do not pose any greater problems to landfills than do other types of wastes such as construction or demolition wastes and newspapers, all of which occupy enormous amounts of space. Since plastics are resistant to decomposition and degradation, these planners contend that plastics do not contribute to leachate and methane explosions. They also maintain that plastics lend stability to landfills as they settle, helping to support the structural integrity of the final cover. Moreover, they note, although other substances, such as paper, eventually decompose while plastics do not, the decomposition of paper is not rapid enough to significantly prolong the life span of a landfill.[5] It must be noted, however, that most modern landfill planners are not necessarily managing a landfill as a composter or digester, and they are not seeking to maximize breakdown for space maximization or methane recovery.

Although plastics themselves are not degradable, questions have been raised about the possible leaching of plasticizers, compounds added to plastics to give them special properties (see Appendix C). Of particular concern is the possible leaching of plasticizers from PVC packaging. Residue from materials known to be used in plasticizers, such as lead, has been found in monitoring wells and in groundwater samples associated with landfills. However, there is as yet no proof that plasticizers do or do not leach. (For more information on the degradability of plastics, see chapter 4.)

Given the increasing prevalence of plastics in municipal landfills, scientific investigations and research on the topic have been slight thus far. While plastics seem to be inert and are projected to be so for thousands of years, these assumptions have not been tested in real-life conditions.

Resource Recovery

The waste option most often advanced as the alternative to landfilling is modern incineration, or resource recovery, so-named because energy recovered through combustion is converted to steam or electric power. As landfills fill up and are closed, and as sites for new ones become more difficult to find, resource recovery will become ever a stronger option.

Resource-recovery plants are not new; they have operated in Europe, Japan, and the United States for many years. However, they are not without potential environmental hazards, both from stack gas emissions and from landfilled bottom ash and pollution control residues containing toxics and heavy metals left over or removed during the process. Operations of plants, the use of air pollution control devices, and the composition of the waste can all influence the efficiency of burning, which, in turn, helps to control harmful emissions.

Of particular concern is the safety of incinerating increasing amounts of plastic packaging containing a large mixture of different plastic resins, laminates, plasticizers, and additives. Chlorinated plastics, such as PVC, certainly contribute to one known problem in incineration: the formulation of corrosive hydrochloric acid gases. Chlorinated plastics may also be part of a mix of substances containing chlorine that form a precursor to dioxin in the burning process. So, although plastic waste has a high energy content and burns efficiently, leaving little residue, there are still many unanswered questions about the effects of burning it, especially in combination with many other materials.

Major questions surround the incineration of plastic wastes, including:

- Does incineration of waste containing plastics result in increased emissions of hazardous materials, directly attributable to the presence of those plastics wastes?
- What is the contribution of other waste-stream components to these emissions?

- What is the relationship of waste composition as a whole to emissions?
- Could certain emissions be reduced if specific materials — including plastics — were removed from the waste stream prior to incineration?
- What, if any, is the contribution of plastic waste to the formation of dioxin/furan emissions?
- Are there specific hazards associated with incinerating waste containing chlorinated plastics?

Resource recovery can mean a number of systems that have been developed over recent years, including mass-burn incineration, refuse-derived fuel (prepared fuel in a process that separates many nonburnable items), and pyrolysis. However, in general discussion, *resource recovery* has come to mean mass-burn, or an incinerator that burns virtually all waste brought to it, at a very high heat, destroying approximately 85 percent of the volume of the wastes. While other systems have their champions and may actually be further developed in the future, most of the plants now being constructed or planned are of the mass-burn type. Mass-burn systems welcome all plastics as high-energy materials, with the possible exception of PVC.

All mass-burn facilities use similar systems. The combined waste is brought to a storage pit or tipping floor; a conveyor system of some type transfers the wastes to the hopper; special grates regularly bring the wastes to the furnace; a combustion air system and auxiliary burners aid complete combustion; and air pollution equipment, ash quench and removal systems, and a high stack control the emission of waste gases. A temperature of 1800°F is considered the optimum for the destruction of pollutants. Therefore, careful operation of such a facility is crucial.

Unlike the older incinerators now being phased out, modern mass-burn plants recover heat from the combustion process through a water-wall of pipes. The water in the pipes turns to steam, which is used for electricity generation or process steam.

Usually, bottom ash and fly ash residue from the control system are combined and landfilled. Fly ash consists of particulate matter captured from the flue gases by the air pollution control equipment. Bottom ash consists of the residue from the combustion system and includes unburnables that were in the original combined waste. Since fly ash is certainly toxic and bottom ash likely to be less so, a controversy has developed as to the proper way to manage these substances. While industry and most government clients prefer to mix the two residues (thus helping to "detoxify" fly ash by dilution) and to treat the resulting mixture as "special waste," environmentalists and others argue for the separation of the two substances, the testing of both, and the landfilling of either in a hazardous waste landfill, if necessary. Such landfilling, of course, considerably increases the costs of mass-burn systems.

Emissions from resource-recovery mass-burn plants may include acid gases, oxides of nitrogen, particulates, and heavy metals, as well as products of incomplete combustion and trace organic compounds. The pollutants of greatest concern are organic compounds known as polychlorinated dibenzo-p-dioxins (PCDDs) and polychlorinated dibenzofurans (PCDFs). Although many fractions of the waste stream provide precursor elements for the production of these pollutants, the presence of plastics has created the most controversy. Claims and counterclaims have dominated both public discussion and scientific investigations. Some plastics do, indeed, contain chlorine, which may provide precursors for the production of PCDDs and PCDFs; but other sources of such precursors include textiles, rubber, yard wastes, and paper.

To date, very few studies of burning have been performed that could connect isolated plastics to the formation of any particular pollutants in the combustion process. The exception, as noted, is PVC. One of the first studies was performed by two chemical engineers from New York University, Elmer Kaiser and Arrigo Carotti, who studied the incineration of municipal solid refuse in 1971. Kaiser and Carotti added polyethylene, PS, polyurethane, and PVC to mixed waste, using the incinerator in Babylon, New York. Another study done by Kaiser and Carotti that same year studied the effects of adding 2 percent and 4 percent of PET to the municipal refuse.[6]

The questions Kaiser and Carotti sought to answer included:

- Will the plastics melt and clog the grates, or drop through the grate openings, carrying fire into the ash pits?
- Will the higher calorific value of plastics cause excessive incinerator temperatures and furnace damage?
- Will plastics increase the air pollution from incinerators?
- Will the presence of plastics in refuse aid in the combustion of wet refuse?
- Will the higher calorific value of plastics facilitate the use of the incineration process to generate steam?
- What improvements in incinerator design and operation will make the burning of plastics more efficient?

In Kaiser and Carotti's test, mixed refuse, which normally contained much less plastic at that time, was incinerated alone as a reference, followed by the combustion of refuse with 2 percent and 4 percent plastic added to it. It is important to note that the tests used commercial products rather than pure polymers. Since plastics are generally compounded into formulations that include pigments, plasticizers, stabilizers, and other agents, it was decided to work with the actual products that might be in municipal refuse.

All tests were performed under normal operating conditions. The data showed that, since plastics have calorific values higher than that of normal refuse, they provide heat that assists in the thorough burning of wet refuse. Plastics were found to burn readily and not drip through the grates. The use of scrubbers in pollution control was recommended to mitigate the acid gas produced by PVCs.

Specific results included:

- The addition of polyethylene did not cause a "significant difference" in smoke density and improved the combustion of the waste, resulting in a decrease in the concentration of organic acids. There was an increase in the nitric oxide concentration attributed to high combustion temperatures.

- PS burned cleanly. No smoke was detected, no melted plastic formed in the grate siftings, and no clogging of the grates by plastic was observed. The high calorific value of PS was found to aid in the burning of wet refuse. Since PS is lower in sulfur than other refuse, there was a decrease in the concentration of sulfur dioxide in the flue gas. The concentration of organic acids in the flue gas increased with the addition of PS, which was attributed to inadequate turbulence in the furnace.

- The odor of the flue gas increased when PVC was added, as did the concentration of hydrochloric acid gases. Chlorine was also found to attack the grate metal of the furnace. (Subsequent experience with scrubbers has shown that, as they increase the amount of limestone used to control acid gases, large amounts of sludge and solid by-products are created. These by-products have certain disposal problems, which can be expected to increase in direct proportion to the chlorine increase.[7])

- The addition of 4 percent PET to refuse resulted in a slight increase in the odor concentration of flue gas. There was no noted effect on nitric oxide concentrations in the flue gas and no significant changes in concentrations of organic acids or hydrocarbons. There was no smoke. The PET burned readily and seemed to have a slight beneficial effect in hastening refuse burnout. No melted plastic was found in the grate siftings or residue, and no clogging of furnace grates by melted plastic was reported.

In 1983 Industronics, Inc., performed a test burn on PET bottles in a pilot waste-to-energy system located in South Windsor, Connecticut. Baled whole PET bottles – including basecups and labels – with an energy value of 10,000–12,000 BTU/lb. were used. The burn resulted in 99.9 percent volume reduction and 99.6 percent weight reduction. PET bottles were found to be an excellent energy source; particulate emissions met standards; no acidic gases were formed; and organic constituents were completely destroyed.[8]

Studies conducted for the New York State Energy Research and Development Authority in 1987 focused on determining the relationships between incinerator combustion and operating variables, refuse characteristics, and combustion variables and levels of PCDDs and PCDFs, as well as certain potential precursor compounds in the combustion gases.[9] The study was designed primarily to evaluate the effect of the following parameters on levels of PCDDs/PCDFs in the combustion gases: incinerator operating conditions, PVC content of the waste, moisture content of the waste, and combustion gas variables.

The tests were conducted at the Vicon Resource Recovery Facility in Pittsfield, Massachusetts. The facility has three 120-ton-per-day incinerators with waste heat recovery. Each incinerator has a primary chamber, in which the waste is burned with excess air, and a secondary chamber, in which combustion of the gases is completed. The gases from the secondary chamber are combined in a tertiary duct and then pass to two waste heat boilers operated in parallel. A portion of these gases is recycled to the incinerator, and the remainder passes through a particulate control device and then to the atmosphere through a stack.

Nineteen tests were conducted, consisting of duplicate runs (except in one case) under different test conditions. The test conditions included such variables as four different operating temperatures in the primary chamber (1300°F to 1800°F), four types of waste feeds (normal municipal solid waste, PVC-free waste consisting mainly of cardboard and wood wastes, municipal solid waste added to PVC, and a mix of PVC-free waste combined with PVC) and waste feeds spiked with PVC or water. Not all tests were done at all the varying temperatures and sampling locations.

Combustion gases were monitored at the tertiary duct (after exiting the combustion chambers), at the boiler outlet (before entering the particulate control device), and at the stack. However, not all locations were monitored for each test.

Waste samples were analyzed for PCDDs/PCDFs, total chlorine, and moisture. Incinerator ash samples were analyzed for the percentage of uncombusted material.

The major conclusions of this study were:

- All the levels of PCDDs/PCDFs measured were relatively low compared to most others reported in the literature.
- There was no evidence that the amount of PVC in the waste affected the levels of PCDDs/PCDFs at any of the measurement locations. (Peer Reviewer Marjorie Clarke noted that the study did show some relationship between PVC and PCDF at 1800°F, at boiler outlet and stack. The report did not deem this relationship statistically significant, however, because of the small number of data points.)
- Higher levels of PVC in the waste increased the level of hydrogen chloride in the combustion gases.
- The addition of water to the waste feed had a minimal effect, if any, on the levels of PCDDs and PCDFs.
- PCDD/PCDF and carbon monoxide levels are related: levels of PCDDs/PCDFs increased at high levels of carbon monoxide.
- Incinerator operating temperatures significantly affected the levels of PCDDs/PCDFs at the boiler outlet location; runs at 1300°F showed significantly higher levels of PCDDs/PCDFs than runs at higher temperatures. At the tertiary duct location, low temperature runs showed sig-

nificantly higher levels of PCDDs than the higher temperature runs, but the differences in levels of PCDFs were not statistically significant between the low and high temperature runs.

- Levels of PCDDs/PCDFs were higher in the boiler outlet duct than in the tertiary duct, indicating that formation may occur in the boiler.

- On the average, the PCDD input rates in municipal solid waste were about four times higher than the stack output rates; PCDF output rates were greater than input rates, but only by about 50 percent. (Only levels of air emissions were investigated—levels of PCDDs/PCDFs in the bottom ash and pollution control system fly ash were not measured, nor was the quantity of bottom ash determined.)

- Concentrations of chlorobenzenes and chlorophenol were found to be higher than concentrations of PCDDs/PCDFs. (Marjorie Clarke emphasized that these two families of compounds are dioxin precursors—in other experiments, they have been shown to form dioxins in fly ash.)

In summary, this study showed that the operating temperature, carbon monoxide levels, and levels of PCDDs/PCDFs were interrelated. The levels of these substances are not affected by different waste characteristics. (Marjorie Clarke: "still not sure.") However, research is needed to more accurately characterize the relationship of various sources of chlorine in the waste and emissions of dioxins, furans, and their precursors.

Additional data indicate that hospital incinerators, which contain a high proportion of plastics, are a potential source of PCDDs and PCDFs. The California Air Resources Board evaluated emissions for four days from a hospital refuse incinerator at St. Agnes Medical Center in Fresno, in a test conducted as part of the board's program to assess emissions from stationary sources.[10] The refuse composition was considered to be typical for hospital incinerators: plastic, 30 percent; paper, 65 percent; and other, 5 percent. The moisture count was 10 percent. Sampling for PCDDs and PCDFs was conducted to assess if hospital incinerators could have the potential to emit such compounds. Only two samples were taken—too few to produce convincing evidence—but the results suggested that, even under steady-state incinerator operating conditions, hospital incinerators are a potential source of PCDDs/PCDFs.

A study conducted by the Environmental Protection Agency in 1987 investigated potential problems during transplant conditions that occur when certain solid materials are batch fed into the combustion chamber.[11] Batch feeding can lead to transient overloading conditions, called *puffs*, heavy loadings of products of incomplete combustion (PICs) in the afterburner and gas cleanup system.

When devolatilization occurs rapidly, local oxygen concentrations in the flue gas are almost totally depleted or displaced by the volatilized waste species. A transient puff then moves as a plug through the system. When volatiles are released slowly, they are transferred into the flue gas stream without completely depleting the oxygen. Slow release thus allows more complete oxidation and minimizes puffs.

The study found that polyethylene and PVC burned in equal proportions at 1950°F and 100 percent excess air (considered optimal conditions) produced a variety of unchlorinated and chlorinated toxic organic compounds. Burning either polyethylene or PVC alone resulted in almost no dioxin or furan, and far fewer organic compounds formed. However, dichlorobenzene, a dioxin precursor, was formed when PVC was burned by itself. This synergistic effect of two plastic resins is important in that it more closely resembles the emissions expected from incinerating a diversity of plastics, as normally found in the waste stream.

Although this result may be due in part to the high oxygen demand caused by the burning of highly combustible plastic waste (an effect that may be diluted in larger furnaces where plastics are in lower proportions), it appears likely that PICs could be formed in municipal solid-waste incinerators as a result of transient puffs caused by locally high concentrations of plastics (e.g., bags of fast-food wastes or hospital wastes.) Therefore, waste composition and transient upsets in combustion efficiency, causing puffs of PICs to be emitted, may account for some high levels of dioxin emissions.[12]

It is generally accepted by most scientists and combustion engineers that 1800°F is a sufficiently high temperature to destroy at least 99.99 percent of the most difficult chlorinated compounds such as PCBs, dioxins, and furans, and their precursors. Sufficient oxygen must be supplied and mixed properly with the combustible materials.[13]

Though many researchers in the field agree that design and operation of incinerators to maximize combustion regulation are the best methods of minimizing emissions of toxic organics, the destruction of all dioxins and furans in resource-recovery emissions cannot always be assured by these means alone. Research indicates that further reductions, up to two orders of magnitude, in these emissions can be achieved by means of properly operated emissions control devices, such as a scrubber/baghouse or scrubber/electrostatic precipitator.[14]

While controversy continues about what policies are necessary to ensure the environmentally safe handling of pollution control residues and bottom ash from mass-burn facilities, there are only limited data on the contribution of plastics in wastes to concentrations of heavy metals and other toxic components of ash. Because bottom ash generally consists of incombustible materials, readily burning plastics do not contribute to the volume of bottom ash. However, lead and cadmium may be used as stabilizers in PVC and thus may contribute to ash high in heavy metals. Metals in bottom ash and toxins in residues pose three potential hazards: the danger of leachable metals, the possible release of dust in handling and transporting ash and residue, and the ultimate disposal of both, mixed or unmixed.

Ultimately, responsible policy decisions must be based on

the realization that nothing in the municipal solid-waste steam is "entirely safe" to burn, nor is any one material solely responsible for the emission of toxic substances. With the exception of batteries and PVC, studies to date do not support the theory that removal of any one material, such as PS, would significantly reduce harmful pollutants.

Recycling

Many components of the postconsumer waste stream can be viewed as resources that have significant economic value when recovered. Recovery also conserves natural resources and minimizes environmental impacts associated with waste disposal. A well-run recycling program can reduce the volume of waste going to landfills and incinerators, as well as increase the efficiency of decomposition in the first and combustion in the latter. Recycling is now being recognized as an essential component of an integrated waste-management system.

Since most of the energy required to produce a plastic product goes into the production of feedstock materials, not the manufacturing process, plastic wastes retain most of their original energy content. Thus, producing a plastic product from scrap plastic instead of virgin resin saves approximately 85–90 percent of the energy otherwise used (see table 5).[15]

Incineration does recover part of the BTU energy in plastic wastes by conversion to heat through burning, but incineration does not conserve as much energy as recycling. Landfilling, on the other hand, wastes all of the potential energy.

The primary recycling of plastics, which conserves the highest amount of energy, involves the reconversion of uniform, uncontaminated plastic waste into its original pellet or resin form. The recycled product has physical and chemical characteristics similar to those of the original product. Primary recycling is suitable mainly for industrial wastes, since they gen-

TABLE 5
Energy Used in the Production of Plastic Packaging

End Product	Energy Content (%)	
	Manufacture of Resin	Fabrication of Product
PVC (1/2-gallon container)	85	15
HDPE (1-gallon container)	90	10
LDPE (1-gallon container)	94	6
PS meat tray	83	17

Source: Charles Papke, "Plastic Recycling as a Business," *Resource Recycling,* September/October 1986, pp. 12–13, 32.

erally consist of one type of resin and have not been contaminated by use in consumer products.

Secondary recycling converts waste or scrap plastic into products with characteristics less demanding than the original. This process uses plastic wastes, often postconsumer wastes, that are unsuitable for direct reprocessing due to contamination or nonhomogeneity.

Tertiary recycling converts plastic waste into a fuel for direct energy or for chemicals. Quaternary recycling is the generation of energy from the burning of plastic wastes.[16]

One inescapable point about the recycling of postconsumer waste plastics is that, unlike materials such as metals and glass, plastics cannot be sterilized at present and reused for food packaging. Thus recycled plastic soda bottles do not reappear as bottles, but as fiberfill for jackets and pillows, for example. While it is desirable to recycle as much plastic as possible, most plastics cannot be used in the "closed-loop" recycling so advantageous to other postconsumer materials.

Recycling, especially mixed-plastics recycling, is proposed by many as the solution to waste plastic disposal. Although progress has been made in the past few years, the recycling of postconsumer plastics is limited by the absence of a strong recycling infrastructure. The main obstacles to increased recycling include: the lack of economically feasible collection, separation, and transportation mechanisms; the dearth of large-scale commercial recycling operations capable of handling a heterogeneous mix of contaminated postconsumer materials; and the lack of steady and demanding markets for recycled plastic products.

Recycling of plastic packaging is made more complicated by the requirement of many processsors for a consistent, homogeneous resin supply. Different resins have different chemical structures, which makes them incompatible with one another. But it is difficult to separate the various resins in the packaging waste stream, and many packages are composed of paper, foil and two or three types of plastic resins. (The proliferation of these multilayer structures and coextrusions is discussed in chapter 2.) Another difficulty in recycling plastic packaging is that postconsumer recycling generally lies outside the expertise of the plastics and packaging industries, which have little knowledge or experience in recycling the products they produce.

To date, most efforts have concentrated on the reclamation of PET and HDPE containers. Virtually all of the PET recycled comes from collection in states that have returnable beverage-container laws. This provides a source of relatively clean, homogeneous resin, creating an impetus for development of end markets. End products are primarily polyester fibers.

Recycling of HDPE occurs as a result of scattered community recycling programs. Uncontaminated HDPE bottles are a good source of recyclable plastic, and nonfood containers can be recycled into similar containers. End products also include drainage pipes and tubes, lumber, flowerpots, toys, and basecups for soft-drink containers.

Mixed postconsumer plastics are easier to recycle since presorting by resins is unnecessary. High process costs and undeveloped and unproven markets for mixed plastic items in the United States has meant that there are few community recycling programs for mixed plastics in existence. Several European technologies for processing mixed-plastic waste are now operating at industrial scale here and abroad.

Many plastics fabricating operations result in a significant amount of waste; manufacturers, therefore, have traditionally recycled their process scrap. Resin suppliers and processors reuse between 3 and 5 billion pounds of manufacturing waste, including "off-spec" resins and scrap generated by independent processors.[17] The material may be recycled back into the fabrication process and made into a product, or the material is collected and sold to another processor for use. Nearly 75 percent of waste plastics generated during fabrication is reused.[18]

As previously mentioned, the most successful postconsumer recycling to date involves PET beverage containers. A typical 2-liter PET bottle weighs approximately 91 grams, 70 percent of which is PET, 24 percent the HDPE basecup, 5 percent the label and adhesive, and 1 percent the aluminum cap.[19] The PET recycling technologies developed for commercial use can economically separate these components, based on their different physical properties.

PET recycling processes vary, but the basic operation includes the separation of the HDPE basecup, aluminum, and paper, resulting in a clean PET flake that can be put to a number of end uses. In dry reclamation systems, HDPE basecups and aluminum closures are mechanically separated. Air separation may also be used to remove the paper and separate the lighter basecup material and the aluminum. PET bottles are then separated by color and fed into grinders. The basecups are ground separately and the aluminum is collected for resale. Postseparation washing is still required to assure the highest sales value.

In wet reclamation systems, the whole bottle is fed into the grinder in one operation. After the grinding, the pieces of the paper label are removed by air filtration. The remaining material is separated by water flotation: PET and aluminum will sink, while the lighter HDPE material floats to the top to be removed. PET and aluminum are dewatered and dried; the aluminum is then separated electrostatically.[20]

Although there are increasing pressures on the industry to be able to "close the loop" for PET recycling, producing new bottles from old ones is unfeasible at present. The resin cannot be reprocessed at high enough temperatures to gain Food and Drug Administration (FDA) acceptance for food-contact applications; the required temperatures would degrade the resin. In addition, the industry fears that the liability risks from using recycled PET in food-contact materials would be too great.

About 75 percent of recycled PET is reprocessed for use in fiberfill or strapping; fiberfill is used in ski jackets, pillows, and sleeping bags. The other 25 percent is reprocessed and blended into compounds with various additives or fillers. The vast majority of reprocessed PET is converted into end markets by Wellman, Inc., in Johnsonville, South Carolina. Wellman processes approximately 100 million pounds of PET annually.[21]

The amount of fiberfill required for a man's ski jacket can be made from five 2-liter PET bottles; a sleeping bag requires 36 bottles. Other uses include unsaturated polyester molding compounds for sinks, shower stalls, corrugated awnings, and exterior panels for automobiles; polyols for making foam used in home and commercial freezer insulation; furniture cushions; paint and industrial coatings; road-building material; and in housings and casings.[22]

According to officials at Wellman, there are "abundant" end markets for products made from recycled PET in the carpet and apparel industries, and in automotive, geotextile, and home furnishings. The main constraint is said to be the limited supply, due to a shortage of collection systems. It is estimated that the market potential for recycled PET in the U.S. exceeds 1.3 billion pounds a year, ten times the amount currently collected for recycling, and almost twice the amount of PET used by the beverage-bottle industry.[23] However, some smaller PET processors indicate that end markets are inadequate and would need to be developed if PET recycling increases. There was general agreement among most processors of PET that green PET was more difficult to market than clear.[24]

The fact that almost all PET being recycled comes from states with bottle-deposit legislation illustrates the lack of effective collection mechanisms for this most recyclable and collectable postconsumer plastic package. PET resin suppliers, as well as soft-drink industry leaders, have begun to commit time and money for PET recycling programs and public education. It is hoped that these actions will expand PET recycling into all states and result in development of expanded markets for recycled PET.

Commercially available recycling technologies and end markets exist for HDPE, but the true potential for its recycling has not been realized. Nearly 2.2 billion pounds of HDPE household bottles were sold in 1987, three times the amount of PET sold. However, only 1 percent of the HDPE bottles were recycled. At present there are few organized community recycling programs for HDPE and, as a result, a scant recycling infrastructure exists.

Many communities, as they plan mandated recycling programs, are beginning to include HDPE (as well as PET) in their collection efforts. HDPE processors, such as Midwest Plastics in Chicago and Eaglebrook Plastics in Chicago, have a process that can tolerate typical postconsumer contamination up to 2–3 percent, including product residue, paper labels, and slight dirt.

Processing postconsumer HDPE involves separating out contaminants, magnetic separation of ferrous contaminants, grinding of the HDPE, and removal of particles. Further processing separates the remaining paper, and washes and dries the purified HDPE.[25]

According to Tom Tomaszek of Eaglebrook Plastics (currently with Polystyrene Recycling Corporation), the major obstacles to increased HDPE recycling are collection difficulties, not lack of end markets. Because one cubic yard of bottles weighs approximately 25 pounds, transporting whole bottles is cost-prohibitive for all but the few communities close to recycling operations. Baling and shredding of the bottles may lower transportation costs, but this equipment is expensive for communities to purchase.

Recycled HDPE is used in producing plastic lumber for boat piers, pig and calf pens, fence posts, and garden furniture; basecups for soft-drink bottles; flowerpots; pipe and drainage tiles; toys; traffic barrier cones; and trash cans. Although FDA limitations apply on food packages, recycled HDPE could be used for packaging of nonfood products such as laundry detergent.[26]

According to Mr. Tomaszek, even if HDPE recycling becomes as widespread as glass or aluminum recycling, the industry will be able to consume the supply and end markets are plentiful. Colorless HDPE receives the highest price, since it can be colored for any end market; mixed colors receive the lowest.

Eaglebrook Plastics processes both industrial and postconsumer waste HDPE. Approximately 40 percent of their material is postconsumer, which the company says it is "aggressively" seeking from "various sources." Eaglebrook processes HDPE from dairy, detergent, motor oil, and bleach bottles; most is reprocessed into irrigation pipes and plastic lumber. A new company formed by Eaglebrook — Eaglebrook Profiles — produces and markets its plastic lumber.[27]

Although containers are not marked or coded, PET and HDPE are easy to identify and can be recovered in homogeneous form. Generally, separation is achieved according to type of container (soda bottles for PET; laundry, milk, water bottles for HDPE), not resin type. Coding could completely alleviate any misidentification.

For the most part, recycling of single resins of other types of postconsumer plastic is non-existent. Based on listings in the Society for the Plastics Industry's (SPI) *Plastic Bottle Recycling Directory and Reference Guide*, there are many companies that recycle other resins in addition to PET and HDPE. However, most of these companies handle plant scrap or postconsumer plastic waste that is segregated, such as PP battery cases or PVC cable scrap. Postconsumer PP is reclaimed and recycled from auto battery cases into a variety of injection molded industry parts. PVC is recovered from cable scrap and recycled for use, either for new wiring insulation or for molding and extruding small plastic parts. With respect to plastic packaging, there is virtually no recycling of resins other than PET and HDPE.

One such company seeking to expand to many postconsumer products is Resources Recovery/Borough Bronx (R2B2) in the Bronx, New York City. Another, Vermont Republic Industries (VRI), of St. Albans, Vermont, receives PVC computer chip packaging from computer manufacturing companies. This is a nonprofit sheltered workshop for people with disabilities. The labels are removed by hand, plugs by machine. They currently process 320,000 pounds of PVC a year; clean PVC is sold to pipe extruders. According to Al Voegele of VRI, this is a very large market.[28] It is hoped that, gradually, these small operations will expand and will become better known through industry sponsorship of information.

There is currently no large-scale recycling of postconsumer LDPE in the United States except for new pilot efforts to recycle plastic grocery sacks. Three hundred million pounds of polyethylene film scrap (mostly LDPE) is currently recovered for use in trash bags. This scrap is supplied by extruders and converters, who make merchandise and specialty bags. They collect, bale, and ship the scrap to processors. This is economical for them, since they would otherwise need to pay for it to be landfilled. The scrap material is blended with virgin resin. Because color is not critical for trash bags and the use does not entail food contact, trash bags are suitable for reprocessed film.

Plastic grocery bags made of LDPE are a potential major source of recyclable materials if the problems of collection, transportation, and contamination from ink and additives can be overcome.[29]

Although PS plant scrap is routinely recycled within manufacturing plants, at present there is no large-scale recycling of postconsumer PS. Plant scrap is reextruded into pellet form and reused for production. (See chapter 2 for details of McDonalds/ Amoco Foam planned pilot program for PS recycling.)

MIXED-PLASTICS RECYCLING

Postconsumer mixed-plastic waste is a heterogeneous mixture of different resins, as well as paper, aluminum, and other contaminants that are often difficult to separate. Because the different polymers are incompatible and do not adhere to one another, mixed polymers can be made only into products that have wide tolerance specifications.[30]

Several European technologies are capable of processing postconsumer mixed-plastic wastes. These processes involve the manufacture of plastic products from a relatively contaminated, heterogeneous feedstock. Products made from mixed plastics include lumber and piping. Advanced Recycling Technology of Belgium has developed and patented the ET/1 (Extruder Technology 1) system, and over 20 plants now operate around the world. In the United States, there are plants at Rutgers University's Plastic Recycling Research Center in New Jersey and at several Long Island locations. These plants are capable of using both industrial and postconsumer scrap as feedstock.

The ET/1 accepts highly contaminated mixed-waste plastics in a typical proportion of 60 percent polyolefins and 40 percent other plastics and nonplastic materials, which is the approximate proportion found in postconsumer plastic wastes. The polyolefins soften to become the carrier, while other materials become fillers that lend rigidity to the end products. Organic contaminants present in commercial or municipal wastes are destroyed by the high temperature in the extruder. The ET/1 can also adjust to other "recipes" and melt ranges. PVC is best limited only to 10 percent or less by weight; however, with the addition of a thermal stabilizer, the mix can tolerate up to 50 percent PVC. PET, which acts as a filler and adds stiffness to the end product, should not exceed 10 percent by weight.[31]

Incoming waste is shredded and sometimes washed. It is reduced in size to pass through a screen with perforations of approximately 8 millimeters. The material, usually a blend of wastes from different sources, is homogenized in a vertical auger mixer, and additives are sometimes blended in. The mixture is then discharged into an intermediate hopper, passed over a magnetic separator, and carried by a special conveyor into the hopper of the extruder. In the extruder, a hydraulically driven high-speed adiabatic screw heats the mix by friction to 200°–300°C and then feeds the melt into the molds. The molds are mounted on a rotating turret and successively presented to the extruder for filling. They are externally cooled by a circulated waterbath and then stored flat for 8 to 10 hours while the centers cool and the products stabilize.

The end product of the ET/1 process is a woodlike product molded into poles, stakes, and slats of various dimensions. Applications include boardwalks, piling and staging in marshlands, dock surfaces and piers, pigsty floor slats, and other agricultural and marine uses. Manufacturers of ET/1 claim that it is water resistant; rot and bacteria proof; resistant to salt water, chemicals, and urine; and will not splinter or split.[32] These products have not been in use for very long, however, and laboratory tests are still being conducted to determine their specific properties.

Another mixed-plastic recycling technology is from Recycloplast, of West Germany, established in 1984. The company's patented process can operate on highly contaminated household waste plastics, as well as contaminated industrial scrap. Recycloplast products are made from batches containing 50–70 percent thermoplastics and 30–50 percent other materials (thermosets, metals, wood, paper, grit, etc.), which may be contaminants present in the waste stream or fillers added to improve the quality of the end product.

The typical mixture consists of 60–65 percent polyolefins.

The mixture is blended and softened by friction heat (200°C). The polyolefins become the carrier material, with the higher-melt-range plastics and nonplastic materials suspended in the mixture. The Replast is extruded in loaves, conveyed to hydraulic presses, and molded into end products. The presses are glass-encased and air filtration devices throughout the plant draw off emissions to a gas scrubber system. This pollution control is designed to prevent PVC in the mixture from releasing hydrochloric acid gas.

Replast products are typically large, bulky items, including cable reels, grates, pallets, sign bases, manhole covers, sound-absorbing walls, and compost boxes. Applications are usually ones where color and appearance are less important than strength, durability, and price. These items mainly substitute for concrete, wood, and metals.[33]

Several plants operating in Europe rely on a new West German technology to recover polyolefin film from packaging and industrial wastes. The process is reportedly able to produce very clean recycled film flake even from dirty film scrap. The feedstock is in the form of baled film, with the quality and reusability of the recovered material depending on the feedstocks; for example, semitransparent films cannot be produced from colored scrap. The process is said to be capable of producing recycled materials with physical properties essentially identical to virgin films.[34]

Selected Pilot Programs

Pilot programs are now under way to test the feasibility of the collection and recycling of postconsumer plastics. The pilots are a necessary first step in full-scale plastics recycling. Today, the vast majority of these programs target PET and HDPE.

NEW JERSEY

As part of New Jersey's mandatory recycling act, plastic and bimetal beverage containers were required to achieve a recycling rate equal to that of glass by April 1988. Beginning in September 1987, manufacturers of plastic beverage bottles announced a multimillion dollar program for the collection, processing, and marketing of PET containers. The program offered funding for collection and processing equipment for municipal and county programs, as well as a guaranteed market for the separated materials. Success, however, is also contingent on the expansion of the program to include other containers, such as HDPE milk, juice, and mineral water bottles. Subsequently, legislative initiatives were introduced in the state legislature to control the proliferation of plastic packaging and disposables. These proposals ranged from banning plastic disposables to imposing a tax on "nondegradable," nonrecyclable "single-use" containers. This plethora of initiatives came largely from local pressures as alarm over plastics became widespread.[35]

Several postconsumer plastics recycling programs have been set up in New Jersey. For example, the town of Mt. Olive, in Morris County, collects PET and HDPE containers along with other recyclables as part of their curbside recycling program. By May 1988 over 30 tons of plastics had been collected since the program's inception in 1987. The collected plastic is transported to the Center for Plastics Recyling Research (CPRR) at Rutgers University. Rutgers does not pay Mt. Olive for the material, but it provides a trailer and a pickup service, as well as a consistent, reliable processing market. Although the Mt. Olive program demonstrates that plastics can be collected and transported, it is important to note that the quality demands of CPRR may not be as stringent as other markets, since the Rutgers program is a research, not a commercial, venture. Commercial recyclers may require that the materials be baled, or that lids be removed, or that the material be fully homogeneous.

Although Mt. Olive residents expressed some initial confusion about exactly what to recycle, their questions were answered and the participation rate is approximately 80 percent. The recycling program has resulted in the saving of landfill space and costs (tipping costs in May 1988 were $113/ton).[36]

Several other communities in New Jersey are also collecting mixed plastics for CPRR: Berlin and Clementon townships in Camden County (curbside) and North Brunswick, Metuchen, South Plainfield, and Piscataway in Middlesex County (drop-off). Four counties—Bergen, Hunterdon, Sussex, and Warren—have included plastic containers in recycling plans submitted to the state. Pilot programs are planned by Atlantic, Burlington, Camden, and Somerset counties.[37]

CHARLOTTE, NORTH CAROLINA

Mecklenburg County launched a voluntary multimaterial curbside collection program for Charlotte in February 1987. The program started with one truck and 2,500 households and has grown ever since. The residents were given 1.5-cubic-foot containers and asked to place commingled PET, cans, and glass bottles in the container and stacked newspapers on top. Overall participation was found to be 80 percent. The rate of set-out varied by material: newspaper, 92 percent; glass, 63 percent; PET, 53 percent; aluminum or bimetal cans, 48 percent; nontargeted materials, 6 percent. The PET collected in this program is processed by Wellman, Inc., as well as by other processors of PET.[38]

RHODE ISLAND

Rhode Island's Flow Control Law of 1986 established an integrated solid-waste system and authorized the Rhode Island Recycling Program, with a goal of recycling a minimum of 15 percent of all solid waste.

A pilot curbside recycling program involving 4,500 households in two towns proved very successful—the main problem was that the trucks filled up so fast that more trips to the materials recovery facility were needed. In East Greenwich, 2,025 households yielded 10,866 pounds of PET and 6,699 pounds of HDPE in 26 weeks. West Warwick's 2,734 households yielded 14,114 pounds of PET and 8,180 pounds of HDPE.[39] Markets for collected materials include Wellman, Inc., and Eaglebrook Plastics.

In Rhode Island's study of its pilot, plastics were found to comprise 3.7 percent of the weight, but 35.5 percent of the volume, of the recyclables collected. PET was found to be 40 pounds per cubic yard of landfill space and HDPE to be 24 pounds per cubic yard. Although only PET and HDPE were targeted for collection, 18.3 percent of the residue collected in the pilot program was plastic other than soda bottles and milk jugs. The state planned to begin collecting mixed-plastic containers in spring 1989 to determine how best to collect, sort, process, and market mixed-plastic containers.

A preliminary evaluation of the pilot program concluded that the following strategies would improve the cost-effectiveness of recycling plastics: increasing truck productivity and capacity; decreasing volumes by grinding the material; increasing marketing efforts; improving processing operations; improving data collection methods; and disseminating more information to the public.[40]

MASSACHUSETTS

In 1988 Massachusetts outlined a five-year plan to add household plastic wastes to its statewide multimaterial recycling program and to reduce the solid waste stream by 25 percent through recycling and composting. The recycling system calls for the curbside collection of materials and the processing and marketing of these materials at recovery facilities.

Two pilot programs will involve the collection of rigid plastic containers and plastic film. A recovery rate of 45 percent of rigid plastics is anticipated. Both the collection trucks and the materials recovery facility will be designed to handle plastics. Several plastic recycling technologies will be pursued, including a polyolefin separation plant and a mixed-plastics plant.[41]

In its study, Massachusetts found that existing markets for PET and HDPE regrind were growing, while markets for polyolefins had growth potential. However, it was determined that markets for lumberlike products would need to be developed, and government procurement programs were seen as playing an important role in establishing these markets. Given the goals of the state plan, it is anticipated that the nation's first full-scale plastics recycling industry could be developed in Massachusetts.

NEW YORK CITY

New York City's recycling projects to date have not focused on plastics; rather, the city's programs target newspapers, glass, metals, and high-grade paper. Through city sponsorship of projects operated outside the Department of Sanitation, however, some small amounts of plastics are being collected, processed, and marketed by Resources Recovery/Borough Bronx (R2B2).

The Environmental Action Coalition (EAC) program for multifamily dwellings, focusing on apartment buildings of over 50 units, has developed a successful strategy for collecting HDPE detergent, milk, and water jugs. Although no effort is made to collect PET beverage containers (since they are collected by retail outlets under the state's returnable container law), some PET has, nevertheless, been recovered by the EAC program. In January 1989, for example, 25 of EAC's organized buildings collected 528 pounds of primarily HDPE. Given the Rhode Island ratio of lbs/cubic yard, this roughly translates to 22 cubic yards of saved landfill space.

The Village Green Recycling Team, an active voluntary drop-off center in Greenwich Village operating since the early 1970s, also collects a variety of plastics. Village Green has encouraged recyclers to bring in all plastics, although sorting has proved to be difficult, due to lack of codes on packages.

Both EAC and the Village Green market their collected plastics to R2B2. Currently, it is doubtful that either program could collect plastics if it were not for the cooperative program with this particular intermediate market. R2B2 is a multimaterial buy-back center that opened in March 1982 and began phasing in postconsumer plastics in March 1983. R2B2 purchases plastics from institutions, commercial enterprises, and the public at large, as well as cooperating with programs described earlier.

R2B2 runs one of the most diverse and aggressive marketing operations in the nation, constantly expanding its capacity to collect and preprocess plastics of all kinds. In 1989 R2B2 processed approximately 600 tons of plastics; at full projected capacity, the annual rate will approach 16,000 tons of separated plastics. Although R2B2 has handled small amounts of mixed plastics, the emphasis is clearly on collections of homogeneous materials for top market prices—a decision that, in part, reflects research and discussions with EAC and others about ways to stimulate successful plastics recycling for the highest end uses.

R2B2 is also planning to open additional buy-back centers in New York City and has been active in replicating its model in other parts of the country. A creation of Bronx 2000, a local development corporation, R2B2 is supported by revenues from its marketing and by other funds, including a Department of Sanitation contract. Since the Department of Sanitation also partially funds EAC and donates to the Village Green, plastics collections in New York City are, at present, substantially subsidized.

Research Centers

THE CENTER FOR PLASTICS RECYCLING RESEARCH (CPRR)

In 1984 the Plastic Bottle Institute, a division of the Society of the Plastic Industry, proposed the formation of a nonprofit recycling foundation and a recycling institute to conduct further research on plastics recycling. In 1985 the Plastics Recycling Foundation established the Plastics Recycling Institute at Rutgers, the State University of New Jersey. In October 1986 the Institute was designated as an advanced technology center by the New Jersey Commission on Science and Technology, and its name was changed to the Center for Plastics Recycling Research.

The CPRR has developed a pilot-scale system for the processing of PET and HDPE, including the removal of basecups, caps, liners, labels, and adhesives from soda bottles. It has also acquired the ET/1 process and is researching and piloting the processing of postconsumer mixed-plastic wastes. One field test in early 1989 included park benches made of mixed-plastic lumber.[42]

Industry Initiatives

COUNCIL ON PLASTICS AND PACKAGING IN THE ENVIRONMENT (COPPE)

COPPE is a broad-based coalition of representatives from the plastics packaging, food, and beverage fields formed to develop and disseminate information on the disposal of plastics in solid waste. Like most industry initiatives, COPPE was organized in response to the growing number of legislative initiatives and public pressures perceived to be inimical to the interests of the plastics and plastic packaging companies.

COPPE's steering committee members include Campbell's Food Company, Coca-Cola, Dow Chemical, DuPont, the Flexible Packaging Association, the National Soft Drink Association, Owens-Illinois, Procter and Gamble, and Shell Chemical.

THE PLASTICS RECYCLING CORPORATION OF NEW JERSEY (PRCNJ)

PRCNJ was formed to promote the recycling of plastic beverage containers, especially in New Jersey. It is funded by members of the New Jersey Soft Drink Association and the National Association for Plastic Container Recovery (NAPCOR). PRCNJ's goals include increasing the recycling of plastic beverage containers, assuring a market for New Jersey containers, providing transportation assistance, assisting localities in purchasing and installing equipment, and working with New Jersey legislators and the executive branch of state government to assure a statewide plastics recycling program that will meet New Jersey's recycling law.[43]

Despite the increased activity of PRCNJ, the milk, juice, and bottled water industries in New Jersey are not yet committed to recycling.

NATIONAL ASSOCIATION FOR PLASTIC CONTAINER RECOVERY (NAPCOR)

In response to the proliferation of container recycling laws and other restrictive packaging legislation around the country, nine U.S. PET resin producers and bottle manufacturers formed a new trade association, NAPCOR. The group plans to work with state and local governments to encourage voluntary recycling efforts. NAPCOR provides financial, technical, informational, and promotional assistance for PET recycling activities, including public education. NAPCOR's members include DuPont, Amoco Container, the Chemical Division of Eastman Kodak

Company, ICI Americas, the Goodyear Tire and Rubber's Polyester Division.

For 1988 NAPCOR had activities planned in nine states:

- Adding PET collection to the existing recycling program
  in Seattle, Washington.
- Underwriting the cost of promotion for the Plastics Re-
  cycling Corporation of California, established to increase
  PET recycling under the California bottle bill.
- Adding PET to the existing curbside program in Austin,
  Texas, and a pilot drop-off program serving five Dallas
  neighborhoods.
- Participating in the evaluation of the feasibility of col-
  lecting PET in Kentucky, through a study conducted by
  the Beverage Industry Recycling Program.
- Adding PET to existing curbside programs in Minnesota.
- Supporting the existing recycling program in Mecklenburg
  County, North Carolina.
- Adding PET to six existing or planned curbside programs
  in Florida.
- Planning a demonstration in conjunction with the man-
  datory recycling plan in Rhode Island.
- Increasing PET recycling in New Jersey, where NAPCOR
  has three seats on the PRCNJ Board.[44]

THE COUNCIL FOR SOLID WASTE SOLUTIONS

Over a dozen companies have contributed $3 million each over
three years to help solve plastic waste management problems.
The Council will sponsor technical research and public education. It seems committed to a major leadership role in the
entire plastics and packaging allied industries, as evidenced by
the two-day seminar sponsored in New York City in February
1989. At the meeting, the industry began to come to grips with
an expanded view of its responsibilities toward design, production, and eventual disposal of the vast array of materials its

member companies and the allied industrial base produce.

Members include: Dow Chemical, E. I. DuPont de Nemours, Exxon Chemical, Mobil Chemical, Quantum Chemical, Amoco Chemical, and Cain Chemical. The Council operates under the auspices of the Society for the Plastics Industry, but will have independent oversight of most of its programs. Its large resin producer members are aware of the confusion created by the myriad of industry claims to "photodegradability" and "biodegradability," and the Council will emphasize viable recycling rather than encourage this approach. It may seek a complementary strategy on the degradability issue with concerned environmentalists.[45]

Gretchen Brewer, the author of Massachusetts's plastics recycling action plan, considers the formation of the Council to be an "important initiative—the first really serious effort by the plastics industry."

4

Current Controversies

As recognition of the negative effects of proliferating plastic and composite packages has grown, controversies over proper public policies have erupted across all areas of the country. Although the waste-disposal crisis is most acute in the Northeast, other sections of the nation are close behind. Some central and mountain states, while possessing large tracts of land for potential landfills, have, nevertheless, now understood the possible contamination of scarce water supplies. Other states contemplating the building of resource-recovery plants may not be able to develop this method because of air quality standards. Meanwhile, the absence of clear federal policy directives and regulations impedes joint policy-making throughout the states.

Reduction and recycling policies should be of national level and intent. In the past, the production of goods of all kinds rested primarily on a localized base. Food and other perishable commodities circulated within small regions, the amount of packaging of all types was minuscule, and many people had very little need for items produced elsewhere.

Obviously, in modern America, just the reverse is true. Almost all commodities are produced on a national and sometimes international base: food, for example, is shipped thousands of miles to all areas of the nation from warmer sections, such as California and Florida. More and more food is being preprocessed for frozen and "shelf-stable" products. Consumer items from toothpaste to paper products are mass-produced; a visit to any supermarket, drugstore, or department store bears out the similarity of products. Therefore, since decisions on production and packaging are made for the national market, it logically follows that decisions for their reduction and/or recycling should be made at the federal level, specifically through the Environmental Protection Agency (EPA) and the Food and Drug Administration (FDA).

Such is not the case, however, except for directly related health regulations made by the FDA. Precisely at the time when the recyclability of the waste stream changed most (the 1980s), the EPA was absent from the public debate. Only in 1988 did the EPA resume its work in the area of municipal solid waste. Given the vacuum created by EPA's absence, issues pertaining to packaging, waste reduction, recycling, and safe disposal have been debated at local and state levels, creating an ever-escalating political controversy.

Degradability versus Non-degradability

One of plastic's most important properties—its inertness—has now become its most publicly perceived liability. As contrasts have been discussed over differing types of packaging—and given the almost non-existence of plastics recycling—some citizen activists and environmentalists have posed biodegradability as a necessary standard that a package must pass before being declared environmentally desirable. In this debate, paper and paperboard packages have often been advanced as being more degradable, and, therefore, more environmentally desirable. Local pressures that have built around this debate have been a major factor in pushing numerous proposed bans on such materials as polystyrene.

The focus on degradability began several years ago when

many state and local initiatives mandated biodegradable loops
for six-packs of beverage containers. These laws, which were a
response to littering, had as their original intent a return to the
cardboard beverage-container holders. Much of the intent centered around the statistical data that showed the dangers of
plastic loops to certain wildlife, which are strangled when loops
become entangled around their necks. One such law is the New
York State Returnable Container Law that mandates the degradability of six-pack loops.

However, since these laws specified degradability rather
than the use of a specific material (such as cardboard), the
plastics companies responded by producing a "photodegradable"
six-pack loop. They claimed that these loops would degrade and
thus disappear from the environment within a short time after
they are littered. But, since six-pack loops form only a tiny
portion of the waste stream, no major focus was put on following up these claims or on getting independent verification of
the claims of "photodegradability."

More recently, the proliferation of fast-food plastic packaging and the replacement of paper grocery bags with plastic
bags made high-visibility packaging the focus of public concern.
These factors coincided with the public's concerns over the
proposals to build increased numbers of new resource-recovery
plants and a push for recycling as an alternative to more incinerators and landfills. The "degradability" that had been assumed
for six-pack loops then became part of the debate between consumers and the industry in the controversies over packaging in
general.

There is a major problem in the claims that are now made
by many in the industry for the "biodegradability" or "photodegradability" of packaging. There are no generally accepted, or
legal, definitions of these words; in the absence of definitions,
they are used indiscriminately, perhaps as a way to placate
consumers or to contain the ever-expanding numbers of local
and state bills that restrict packaging.

Biodegradability, for instance, has a biologically accepted

definition—one that presupposes the ability of organic sub-stances to break down into the soil base due to the actions of moisture and oxygen. True biodegradability, of course, is part of the natural ecosystem and an ongoing, essential part of life.

However, "biodegradability" as defined for packaging often means something else. Industry representatives and scientific researchers know that the very properties that make plastics, make them, also, non-degradable. Indeed, that is their functional purpose. Therefore, any plastic that purports to be degradable in any way must be adulterated—and the very inertness that is the essential property of a plastic is damaged.

EAC's research into the meaning of the word "degradable" (on plastic bags in particular) led to a systematic query of a list provided by the Engineering Department of the City of Newark, New Jersey. The list, which Engineer Frank Sudol sent as part of his ongoing search for correct information about plastics, was headed "Some Commercially Available Degradable Plastics." Twelve products were listed; EAC was not able to communicate directly with representatives of all twelve, but had substantial interviews with a large number. EAC also called a number of companies whose names were listed on plastic bags that claimed to be "photodegradable" or "biodegradable." The results were not conclusive, but should give certain pause to those who would seek to mandate degradable plastics in any form.

The results were as follows:

Ampacet Company

Ampacet's product, listed as a "plastic," is actually an additive. Polygrade I, its "photodegradable" additive, is recommended for use with polyethylene, primarily, as 5 percent of the total mix-ture for plastic bags. (As with all companies interviewed, the contents of Ampacet's "photodegradable" additive are proprie-tary.) Polygrade II, a natural starch product, is recommended for use at 15 percent of the total mixture. The company's repre-sentative, Eric Reger, clearly knows that the plastic resin in a bag containing additives does not completely degrade. The additives are designed to adulterate the inert plastic, which then remains as shards or dust after the additives have been released to the environment. While starch products are presumed not to be harmful after breakdown occurs, the effects of the additive for "photodegradable" products are unknown.

Dow Chemical

Georgia Huff, Dow's representative, reported that the company makes "photodegradable" plastic only—not "biodegradable"—which is sold to Owens Illinois Company for use in six-pack loops. The loops are reported to "degrade" in less than a year, depending on the geographic location of their disposal. Dow understands that a covered landfill limits the possibility for degradation of any sort. The rate of "photodegradation" of the loops depends on the amount of ultraviolet light to which they are exposed. The product is now being tested for "verification." Dow will keep EAC informed of the outcome of the tests. The company believes that the final product after breakdown is nontoxic and "will not contaminate the soil." According to the list supplied by Frank Sudol, Dow has applied to the Food and Drug Administration for food-contact use of its "photodegradable" plastic.

DuPont

Company representatives briefly discussed "photodegradable" six-pack loops with EAC staff. The loops, composed of ethylene and carbon dioxide, "degrade" when exposed to ultraviolet light. When the loops break, the carbon dioxide is released, leaving the ethylene in the environment.

Hobbs & Hopkins

This company's product, Saxolin-X Bio-Bag, is not a plastic at all. It is made completely of cellulose and will degrade as a natural product, given moisture and oxygen. It is completely biodegradable, in the classic sense of the word.

Plastigone Company (formerly Ideamasters)

Dr. Robert Ennis reported to EAC that plastigone is actually an additive invented by British and Israeli chemists. It can be used in many plastic resins as a substitute for the stabilizers normally used. The special nature of the product allows plastic to "degrade" in ultraviolet light on a timed basis—on a scale from 3 weeks to 12 or 18 months. Once the "trigger" is tripped, the plastic begins an irreversible degradation—even in the dark. Small pieces of plastic resin, of small molecular weight, degrade into water and carbon. Plastigone is particularly useful, according to Dr. Ennis, in agricultural mulch material.

ICI Americas

According to company representative Tom Galvin, ICI produces PHBV, which is 100 percent resin. It is made by a fermentation process in which naturally recurring sugar is fed to microbes that produce natural polymers, which the company collects and cleans. In 1989 the cost of the resin was a noncompetitive $15 a pound. In 1990, Biopol, ICI's European product being marketed to Wella of West Germany, was becoming economically competitive, with the 1991 price forecast at $6 to $9 a pound (according to an interview with the company's David Barstow in *ChemicalWeek*, May 2, 1990). The time it takes for products with PHBV to degrade depends on the form of process used. In an aerobic sewage system, for example, estimated time is 1 to 2 days. In the ground, degradation might take 3 to 4 months. The product will not degrade in air, being stable, and the byproducts are carbon dioxide and water.

Poly-Tech, Inc.

Poly-Tech's product is a trash bag, sold under the trade names Hi-Flex and Ruffies. It is made of LLDPE, with an additive purchased from an outside company and integrated into the plastic resin with the coloring substances. The company's claim is that the plastic bag should "degrade" within 3 to 10 days, depending on the intensity of the ultraviolet light, the temperature, etc. The residue is "fine, minute particles" of polyethylene, which the company insists are natural carbon products.

At the time that EAC's final interviews on the subject of degradation were taking place, the Newark Municipal Council passed, on February 15, 1989, a local ordinance (#6S&FA) banning the use and sale of PS and PVC food packaging in a number of applications. Part of the ordinance addresses the issue of degradability. "Degradable packaging" is defined as "packaging made of cellulose-based or other substances that are capable of being readily attacked, decomposed, assimilated, and otherwise completely oxidized or broken down by bacteria or other natural biological organisms into carbonaceous soil material or water and carbon dioxide; or in the alternate capable of otherwise degrading within 12 months of manufacture, into fragments that are small relative to the original size, or into particles of a molecular weight that is low when compared to that of the original material."[1] Given EAC's research, this definition, which seeks to clarify the situation surrounding "degradable packaging," is so broad as to encompass virtually all of the claims now proliferating in the plastics industry. Furthermore, this language does not address the potential problems of additives or the final degradation or nondegradation of dust or shards that may remain after the degradation of plastic resins.

It seems certain that controversy will continue at all levels of government on the proposed use of more and more "biodegradable" and "photodegradable" plastics. Ronald Forster, a research scientist at Rutgers University, has undertaken long-range investigations of this issue and will be preparing studies and other scholarly papers in the future. At this point, he has found "no specific studies to show end products and by-products of degradation." Clearly, if policy-makers are to rely on scientific data, there must be a body of work independent from industrial, proprietary claims.[2] In a prepared presentation at the EPA Solid Waste Research and Development Conference in San Diego in January 1989, Richard Renfree, a consulting engineer and colleague of Forster's at Rutgers, stated that plastics are definitely *not* degradable and that more study is needed to assess the ability of plastic resins to complete the natural cycle of degradation that organic materials, such as cellulose, undergo.[3]

In any event, the entire discussion of whether or not products and packages are "biodegradable" or "photodegradable" is almost moot, because of the landfilling methods now in use. If landfills were to be managed as giant composters, selecting and using only those materials in the waste-stream that were truly biodegradable in the accepted biological sense, the separated organic material could yield valuable methane gas fuel and possibly prolong the life of landfills. At present, however, the "sanitary landfill" methods mandated by regulations preclude this type of management. The focus, instead, is on tight compacting and daily cover-up of mixed garbage and trash with clean earthfill. Machines compress the working face even further and an essentially airtight environment is created. Dr. W. L. Rathje, of the Bureau of Applied Research in Anthropology at the University of Arizona, has conducted scientific excavations into modern landfills and recorded data from exhumed materials. He has found a lack of decomposition of organic matter, such as food and newspapers, even over a long period of time.[4] Therefore, policymakers may want to study the conclusions of studies such as Dr. Rathje's before mandating the broadly defined "degradability" of plastics.

True Recyclability and Industry Coding

At a time when recyclability has finally become a highly desired claim, the lack of accepted and legal definitions of terms such as "recycled" and "recyclable" has now become a major problem. Claims of the recyclability of products cannot be independently verified and are essentially meaningless.

Among long-time recyclers—both those in the scrap and used materials industries and not-for-profit activists—"recyclable" generally refers to an item that is fiscally viable to recycle within the general economy of the country and its trading partners. It presupposes that an extensive recycling infrastructure is already in place and that collected and segregated items can enter a regularly functioning system. Using this definition, most plastic products and packages fall far below the recyclability of competitors such as paper, paperboard, metals, and glass.

Even those few plastics—such as PET, HDPE, and LDPE—for which some recycling methodology is in place are difficult to organize for separation and collection because they cannot be distinguished from other plastics. For a number of years, environmentalists and recyclers urged the plastics industry to develop codes that would appear on packages so that they could be identified easily. Initially, the industry resisted these suggestions. During 1988, however, a coding system was developed and offered by the Society for the Plastics Industry. By this time, the controversy over coding had escalated. The "Voluntary Plastic Container Coding System" is a seven-code listing that proposes to cover virtually all plastic packages.

Ironically, in finally responding to the pleas of recyclers, environmentalists, and even government representatives, the industry created a new controversy. The coding system sought to give the impression of complete recyclability by using the time-honored symbol of recycling: three arrows forming a triangular loop. Since most plastic packages are not recyclable in terms of economic viability, the recycling world reacted negatively to the usurption of the recycling symbol by the plastics industry. While some states have mandated the use of the industry coding system on all plastic packages distributed within their borders, others have resisted. Most of the northeastern states, organized under the Council of Northeastern Governors, have rejected the use of the arrows in the suggested codes. They are being supported in this action by recycling and environmental communities. Until this controversy is resolved (perhaps at the national level), progress in the segregation and collection of those few plastics that are developing a recycling infrastructure will be impeded. An example of this problem is the decision by Eaglebrook East Plastics to shut down temporarily its Middletown, New York, facility because the collection of postconsumer HDPE was insufficient to generate the needed scrap. Andrew Stephens, Eaglebrook's president, reported that the plant would reopen when enough scrap is available through state and municipal collection efforts.[5] The plant at printing has never reopened.

Restrictive Legislation

In an effort to reduce and recycle all waste, particularly that relating to all types of packaging, various types of restrictive legislation were put in place by local and state governments during the 1980s. It is significant to note that, in the opinions of most of those in the field, these initiatives were a response to real or perceived crises that were being ignored by the federal government, where, perhaps, some of these initiatives properly belong. It is too soon to tell whether any of these measures will find their way to the national legislative or executive branches.

RETURNABLE BEVERAGE CONTAINER LAWS

By 1988 nine states had passed returnable container laws of some sort, commonly known as bottle bills: New York, Massachusetts, Vermont, Oregon, Michigan, Maine, Delaware, Iowa, and Connecticut. (California's deposit law differs substantially from the others.)

Historically, bottle bill legislation was intended to reduce litter, but it has also resulted in the increased collection of containers, leading to a larger materials base for recycling. No law has yet mandated a return to returnable/refillable bottles, which was one of the original ideas.

The enactment of bottle bills has generally coincided with accelerated replacement of heavier, breakable glass with lighter-weight, unbreakable PET beverage containers. However, metals and glass have continued to keep part of the market, and most studies show that the recyclability of these packages continues at a higher rate than that of PET. In contacts with bottle bill states by EAC in the fall of 1987, it was reported that the return rates for aluminum and glass were in the 90 to 100 percent range and nearly 100 percent of these returns were recycled. PET, which was reported returned at an average rate of 50 percent, was more often landfilled than were the other containers, especially if it was green. For example, in December 1987, all of the PET then being collected in Iowa was being landfilled because of lack of available markets.[6]

New York State's mandatory deposit law on all beer and soft drink containers took effect in September 1983. A study done by the Rockefeller Institute, State University, Albany, after the first year of the implementation reported a 59 percent recycling rate for all cans (including aluminum and bimetal), a 77 percent recycling rate for glass, and a 33 percent rate for plastic. Before the law went into effect, recycling rates in the state were estimated to be 5 percent for cans, 3 percent for glass, and 1 percent for plastics, by weight. An approximate 5 percent reduction in solid waste by weight was reported.[7]

According to a study done for the Aluminum Association by Franklin Associates in 1986, 53 percent of all plastic beverage containers sold in New York State were being recycled into new products. This compared to an 82 percent recycling rate for aluminum cans and a 79 percent recycling rate for nonrefillable glass bottles. The study also found that PET bottle usage in New York increased 15 percent between 1983 and 1984, while aluminum can usage declined 6 percent and nonrefillable glass bottle usage dropped by 60 percent.[8]

The Franklin Associates' study presented data for 1984, the

first full year of the implementation of the deposit law. According to the Bureau of Waste Management of the New York Department of Environmental Conservation, the Franklin data is the most recent information. It is important to note, however, that the data is industry-derived and not independently verified.

PACKAGING TAXES

Essentially, packaging taxes levy a few cents on all packages, with items receiving full or partial relief from the tax based on their proven recyclability and/or secondary materials content. Packaging tax revenues generally are used to fund various solid-waste projects, from landfill closures to research projects in recycling.

The purpose of packaging taxes is to encourage the use of reusable and recyclable materials as well as to encourage the use of recycled materials. Proponents of packaging taxes see them as a way of simplifying the waste stream and making it more manageable, as well as raising revenues for certain solid-waste programs. Critics of packaging taxes claim that the tax is passed on to consumers, that the tax is arbitrary, and that it is difficult to administer.

Many different packaging taxes have been proposed over the past few years, although none has yet become law. One proposal that has received much attention is that introduced in the New York State Assembly in 1988, with a modified version also introduced in 1989.

The original version of the bill would have imposed a three-cent tax on most nonfood and fast-food packaging, with credits and exemptions from the tax based on a package's recyclability and recycled material content. A state packaging review board would have reviewed packages to determine their eligibility for credits upon petition by manufacturers. An official state recycling emblem would have been conferred on packages that met both criteria and were exempt from the tax.

As expected, the bill aroused a great deal of controversy.

Industry groups vehemently opposed it, arguing that it would never work. After hearings the bill underwent several modifications, including the simplifying of the criteria for exemption. The definition of "recyclable" was rewritten to define the word as "being made of one material." As of mid-1990, the bill has yet to become law. However, the recycling emblem that was part of the original bill was passed under the omnibus Solid Waste Act of 1988.

The interest in packaging taxes remains strong, and this type of legislation bears serious consideration at both state and national levels in the future.

MATERIAL BANS

The extreme grass-roots pressure against certain plastics, particularly polystyrene foam, has led to an explosion of local and state initiatives for banning the use of materials in many applications. Most of the controversy has centered on PS used in fast-food packaging and disposable products. Other targets have included packages and utensils made of PVC. The company most visibly targeted has been the McDonald's Corporation, as previously mentioned. Industry has responded with counter-pressures, and in early 1989, the controversy over material bans was the most serious of all the debates over waste-management methods. It seems unlikely that this controversy will abate soon.

Perhaps the best-known ban is that passed by Suffolk County, New York, in March 1988. This law, which was scheduled to go into effect in the summer of 1989, bans items such as PS foam "clamshells," plastic meat trays, and plastic grocery bags. Exemptions include clear plastic used to wrap meat, fish, cheese, cold cuts, produce, or baked goods; packaging used in hospitals or nursing homes; paper or cellulose-based packaging coated with polyethylene on one side only; and plastic containers, covers, lids, or utensils that are not made of PS or PVC. Similar legislation has been introduced elsewhere and was part of Newark's Ordinance #6S&FA passed in February 1989.

A coalition of plastics trade groups filed suit against Suffolk County in an attempt to overturn the law before it went into effect. The lawsuit charges that the State Environmental Quality Review Act was violated, since no comprehensive review of the bill's environmental impact was done. It also argues that the law conflicts with the state's goals of 10 percent waste reduction and 40 percent recycling by 1997, in part because the banned plastic packaging is potentially recyclable, particularly the PS. Because this bill was the first of its kind in the nation, it had been chosen for concentrated opposition; by early 1989, the industry was vigorously engaged in tracking and opposing dozens of similar initiatives across the country.

As with packaging taxes, it seems certain that more ban legislation will be introduced. The vast array of differing details poses special challenges to an industry that is nationally and internationally organized. It is probable that this type of restrictive legislation will be settled only at the federal level in the long run.

FEDERAL ACTION

The Marine Plastic Pollution Research and Control Act, signed into law by President Reagan on December 29, 1987, provides domestic implementing legislation for an international convention that regulates the disposal of all types of garbage from ships, specifically the disposal of plastics. The legislation also contains two provisions specifically relating to plastics. The first requires the EPA to conduct a major analysis of the role of plastics in solid-waste disposal and to analyze methods to reduce plastic through recycling or using other products, with particular attention to reducing land-based sources of plastic debris in the marine environment. The second provision requires the EPA and the National Oceanic and Atmospheric Administration to conduct a public outreach program to educate the public about the problem of plastic pollution.

In 1988 Representative George J. Hochbrueckner (D-N.Y.) introduced the first comprehensive recycling legislation (H.R. 500) in Congress. The bill, the Recyclable Materials Science and Technology Development Act, called for federal research on: technology development for recycling of nondurable consumer product packaging; expansion of markets for recycled products; and encouraging the development of biodegradable consumer products. It also would have established an Office of Recycling Research and Information within the Commerce Department to make grants for recycling research and development. An important part of this initiative was its attempt to encourage the plastics industry to consider recycling as a basic design goal, and it called for scientific research into development of plastics recycling methods and systems, including collection, sorting, reclamation, and end-use manufacturing.

The 1989 Congress expanded the interest of Congressman Hochbrueckner, and it is likely that a number of bills will be introduced in both the House and Senate as part of the reauthorization of the Resource Conservation and Recovery Act. One such bill is the World Environment Policy Act of 1989, sponsored by Senator Albert Gore, Jr. (D-Tenn.), which encompasses major international initiatives to control chemicals and gases harmful to the ozone layer, and also includes language that would phase out, over five years, all nonrecyclable packaging and thereafter prohibit any packaging that cannot be recycled or is not naturally degradable.

Legislation, particularly at the federal level, may be the most effective way to mandate certain disposal and manufacturing practices. However, in order to be effective, new laws must be based on fact, not emotion. Everyone "hates" polystyrene foam, but is it responsible for clogging our landfills? Adequate research *prior* to drafting a bill may eliminate the long delays and redrafts. Information such as waste composition, recycling rates (actual and potential), and the economic and social feasibility of alternative waste-management methods (such as composting and recycling) may result in a more effective and enforceable law.

Beginning with voluntary measures and phasing in mandatory regulations may afford an opportunity to fine-tune the legislation where it is needed and may help industry to change in a more timely way.

5

Conclusions and Questions

The entire question of how to manage the mixed-solid-waste stream is now a topic of intense debate. During the years 1987 and 1988, which have comprised EAC's intensive investigation period, this issue has "exploded" from one in which relatively few people had identified the problem to one in which many people declare that solid-waste disposal questions are "No. 1" on the agenda in many localities and states. In the past six months, this issue has gone to the top of the federal priority list as well.

In 1984 EAC was virtually alone in questioning the true recyclability of postconsumer plastics. By January 1989, the organization had been joined by a nationwide coalition. During this same period, the options for waste management had been turned upside down. From a methodology that had relied on landfills, with attendant resource-recovery incinerators and (perhaps) some recycling, the U.S. Environmental Protection Agency now has declared a national policy that establishes a "hierarchy" of integrated waste management: reduce, recycle, incinerate for waste-to-energy, and landfill.

But will this policy decision make it possible to achieve this laudable goal? Unless major changes take place in the way our packaging is designed, produced, and disposed of, the answer will be a resounding "no."

The following questions constitute the shape of future debates on this issue.

Question 1: *Will the federal government analyze the nationwide production of packaging and products and recommend reduction, both in toxicity and in volume?* Voluntary guidelines, it has been shown, are seldom enough. There is no substitute for clear and concise requirements. This does not necessarily mean bans, although this may be the result in selected cases. Almost certainly, these requirements will involve the consideration of restrictions on materials and combinations of materials. They must certainly involve major decisions on additives and inks. They necessarily will restrict the manufacturer's ability to design "at will" and create packaging without regard to its recyclability and/or safe disposal.

Question 2: *Will the federal government analyze the plethora of local and state measures and organize them into a coherent policy?* Although production in almost every sector of the U.S. economic base is on national and international terms, the regulation of plastics has been accomplished by state-by-state legislation. This has not proved to be the best policy. States do not have the legal power to mandate many crucial policies. Even if legislation survives legal challenges, these laws will not prevent these problems from arising if other states do not follow suit. Given the present void in national policy, the states have had no choice but to put in place or threaten to impose packaging taxes, selected bans, and other restrictive measures. All of these belong at the federal level, as part of an overall waste-management policy for the nation.

Question 3: *Will local, state, and national government come to terms with the institutional barriers that restrict the* *success of recycling?* Examples of this situation abound at all levels. Some examples follow:

- Local building and fire codes prevent the use, design, and rehabilitation of physical spaces for recycling. In some cases, even when the desire and mandate for recycling are present, it is physically impossible to comply, given the restrictions on basements, stairways, etc.
- Local procurement laws not only do not mandate the purchase of recycled goods, they often prevent it by specifications that forbid the inclusion of recycled goods options in bidding. This problem is widespread at all levels of government. There is a great need for government agencies to perform a "pump-primer" role in the creation of expanded markets for recyclables. The lack of demand through large-scale procurement can stymie the proper growth of recycling, just at the time when recycling has become a preferred method of waste management.
- Federal freight rates were long ago fixed to facilitate the transfer of virgin materials as the country developed its hinterlands. Today, these antiquated laws unnecessarily restrict the proper flow of recyclables and recycled materials. Only Congress can remove this institutional impediment.

Question 4: *Will the issue of mandated "biodegradability" and "photodegradability" be opened up for reexamination?* Loose, undefined language in a myriad of local and state laws has created an alarming situation where spurious claims of "degradability" are being made for a host of plastics products. In all cases, plastic remains at the end of any degradation process, no matter whether the additive is a natural product, such as starch, or proprietary additives that react with ultraviolet light to foster degradation.

There is a great absence of clear research and knowledge about the effects of "degradable" plastics on the environment;

it is recommended that no more legislation be passed at any level mandating "biodegradable" or "photodegradable" products until federal definitions have been put in place.

Question 5: *Will a data base of scientific and social statistics be available for informed decision making?* To date, there has been a high concentration on studies of what is currently happening; few projections are available on what will happen in the near and intermediate term. None are available for the long term. The very phrase "integrated waste management" is of recent coinage; when the radical nature of "reduction" is understood, there may be a negative reaction from consumers and, ultimately, decision makers, since reduction of waste means a change in lifestyle. Reduction's place at the top of the waste-management hierarchy may result in changes of a significance that is now unanticipated.

When the Environmental Action Coalition embarked on our packaging review in 1984, our major question was, Is the recyclability of the waste stream at risk?

The answer, as our research clearly demonstrates, is an unequivocal yes. Given the challenge to the nation to dispose of wastes in an environmentally acceptable manner, we must return the waste stream to a state of high recyclability, reduce the toxicity and volume of waste at the source, and make disposal methods as environmentally safe as possible.

Solid waste management and disposal—truly a major agenda for the 1990s.

Appendix A
Manufacturing of Plastics

Plastics are petroleum-based synthetic materials whose main constituents are carbon and hydrogen. In 1868 John Wesley Hyatt mixed pyroxylin (made from cotton and nitric acid) with camphor to create celluloid—the first commercial plastic in the U.S. marketplace. (Hyatt's experiments were undertaken in response to a competition sponsored by a billiard ball manufacturer who was looking for a substitute for ivory.) The first photographic film Eastman used to produce the first motion picture film in 1882 was made of celluloid.

The next major development in the plastics industry took place in 1909, when Dr. Leo Hendrick Baekeland introduced phenolformaldehyde plastics (known generally as *phenolics*). Plastics development continued through the 1920s with the introduction of cellulose acetate, ureaformaldehyde, polyvinyl chloride, and nylon. It was not until the demands of World War II, however, that large-scale production of synthetic materials was spurred to provide a substitute for natural resources in short supply. At this time, plastics began replacing wood, metals, leather, and glass. Polyethylene was a wartime development that grew out of the need for a superior insulating material that could be used for applications such as radar cable. Development accelerated after the war, and continues at an ever-faster pace today.

The essential ingredient of every plastic is a high-molecular-weight polymer, a long chain containing thousands of repeating small molecular units. If various groups are present in the same molecular chain, the polymer is called a *copolymer*. Polymers consisting of repeating units of a single kind are called *homopolymers*.

Most commercial polymers are synthesized from *monomers*, simpler molecules such as ethylene, propylene, benzene, and styrene. The simple chemicals from which monomers and polymers ultimately derive are usually obtained from petroleum and natural gas. About 20 percent of all oil and natural gas consumed in the U.S. each year is used to produce petrochemical feedstocks for the plastics industry (Jacob Leidner, *Plastics Waste, Recovery of Economic Value* [New York: Marcel Dekker, Inc., 1981]).

The most common plastics manufacturing processes are extrusion, thermoforming, injection molding, and blow molding:

Extrusion involves the conversion of plastic powder or granules into a continuous uniform melt, which is forced through a die to yield the desired shape. In *coextrusion* an extrudate of two or more layers, often of different resins, is made by feeding a single die from two or more extruders. Coextrusion produces a laminated product that combines the properties of several resins — properties not available in any one resin.

Thermoforming is a two-step process of elevating the temperature of a thermoplastic material to a workable level and forming the material into the desired shape.

Injection molding feeds a pelletized granular or powdered plastic from a hopper into a heating cylinder. The resulting melt is injected, under high pressure, into a metal mold that gives the finished product a precisely defined shape. Pressure is maintained until the plastic has solidified and is hard enough to be ejected from the mold.

Blow molding uses extrusion, injection, and injection/stretch processes to produce preforms that are then blown into the desired molded shape.

The physical and chemical properties of plastics vary, depending on the type. Melting points range from 80°C to 300°C. Density ranges from 0.92 to over 2. Other variable properties include crystallinity, solubility, burn behavior, and molecular weight. Properties are determined by the linkages between monomers and their structural arrangement, the length and types of molecules in the polymer chain, the integration of differing types of monomers in the same chain, and the incorporation of various chemicals or additives during or after polymerization. Such variations allow plastics to be used in a broad array of applications, but they also make recycling of mixed plastics difficult.

Most plastics fall into two groups, thermoplastics and thermosets:

Thermoplastics, which represent approximately 87 percent of U.S. resin sales, are characterized by their ability to be cooled and hardened and then reprocessed when reheated. Common thermoplastics include polyolefins (polyethylene, polypropylene), styrenes (polystyrene, acrylonitrile butadiene styrene), vinyls (polyvinyl chloride, polyvinylidene chloride), and thermoplastic polyester (polyethylene terephthalate). Thermoplastics are discussed in more detail in Appendix B.

Thermosets represent the remaining 13 percent of U.S. resin sales. Thermosets are infusible and insoluble, and cannot be resoftened by heat. Common thermosets include phenolics, urea-formaldehyde resins, epoxies, and crosslinked polyesters.

Appendix B
Major Types of Thermoplastics

Polyethylene is 85.7 percent carbon and 14.3 percent hydrogen. It is polymerized from ethylene gas at controlled temperatures and pressures. Polyethylene accounts for the largest volume of plastic used in the United States.

Ethylene vinyl acetate (EVA) describes a family of thermoplastic polymers ranging from 5 percent to 50 percent by weight of vinyl acetate incorporated into an ethylene chain. In comparison to polyethylene, EVA copolymers are more permeable to gases and water vapor, have slightly poorer electrical properties and chemical resistance, and are less heat-stable.

Polypropylene (PP) is a stereospecific thermoplastic polymer of propylene with low specific gravity and good resistance to chemicals and fatigue. It consists of 85.7 percent carbon and 14.3 percent hydrogen. It has a specific gravity of 0.90 to 0.91, making it the lightest of the major plastics. Its melting point ranges from 165°C to 170°C. PP can be compounded with fillers or additives to modify its mechanical strength, flammability, and resistance to warping, impact, heat, and ultraviolet light.

Polystyrene (PS) is a versatile thermoplastic made from the petroleum derivatives benzene and ethylene. The two are reacted to form ethylbenzene, which is then cracked to remove one molecule of the monomer styrene. These are then linked to form PS, which is 92.3 percent carbon and 7.7 percent hydrogen.

Polyethylene terephthalate (PET) is prepared from ethylene glycol and either terephthalic acid or the dimethyl ester of this acid. Ethylene glycol uses ethane as feedstock; terephthalic acid uses p-xylene. The melting point of typical commercial PET is approximately 250°C; the melting point of highly crystalline PET is 270°C.

Polyvinyl chloride (PVC) copolymers are based on vinylidene chloride polymerized with methyl acrylate, vinyl chloride, or unsaturated carboxylic acids. Typical formulations contain 2–10 percent plasticizer such as dibutyl sebacate or di-isobutyl adipate and 0.5–1 percent heat stabilizer. The most significant property of these resins is their barrier to gases and liquids; they also exhibit resistance to fats, oils, and many chemicals. The acceptance of a wide range of plasticizers and additives makes PVC one of the most versatile thermoplastics. Chlorine, however, has been linked to corrosive acid gases and is a known dioxin precursor.

Appendix C
Plasticizers and Additives

Plasticizers are "high boiling liquids or low molecular-weight solids that are added to resins to alter their processing and physical properties" (Roy T. Gottesman and Donald Goodman, *Applied Polymer Science*, 2d ed. [American Chemical Society, 1985]). The resulting flexible plasticized resins possess a broad spectrum of additional uses. Commonly used plasticizers include:

Phthalates such as butyl benzyl phthalate, butyl octyl phthalate, diethyl phthalate, diethyl hexyl adipate, di-isodecyl phthalate, di-tridecyl phthalate.

Phosphates such as tricresyl phosphate and tri (2-ethyl hexyl) phosphate.

Trimellitates such as trioctyl trimellitate and tri-iso-octyl trimellitate.

Aliphatic diesters such as dioctyl adipate, n-octyl decyl adipate, and dibutyl sebacate.

Benzoates such as diethylene glycol dibenzoate and dipropylene glycol dibenzoate.

Epoxides such as epoxidized soybean oil and octyl epoxy tallate.

Polymerics in low-, medium-, and high-molecular weights.

The most commonly used plasticizer is diethyl hexylphosphate (DEHP), whose primary use in packaging is for food, principally in PVC packaging film for meat and poultry. There is a possibility that these compounds may migrate from packaging film, depending on temperature, contact surface, and fat content of food (John Wicks, "PVC: The Facts" [Association of Plastics Manufacturers of Europe, 1986]).

DEHP at high concentrations is a recognized carcinogen, with some evidence of mutagenicity and teratogenicity (Wicks, "PVC: The Facts"). Animal experiments to determine the effects of DEHP, however, have been inconclusive. Liver and testicular damage have been reported, but results differed between animals, and results could not always be extrapolated to humans.

Additives, which transform a polymer into a plastic, include colorants, flame retardants, heat or light stabilizers, antioxidants, and lubricants, in addition to plasticizers. Common additives and their uses include:

Antioxidants are organic compounds incorporated at low concentrations to inhibit or retard polymer oxidation and subsequent degradation. They include phenolics, secondary amines, phosphites, and thioesters.

Colorants are used to enhance the appearance of the finished product. They include inorganic pigments such as titanium dioxide (white), iron oxides (yellow, red, and black), chromium oxide (green), lead chromate and molybdates (yellow and orange), cadmium/selenium/mercury compounds (yellow, orange, and red), ultramarine (blue and violet), and a variety of mixed metallic complexes (blues, greens, browns, and violets). Organic pigments are also used. Colorants for food-packaging applications must meet regulations for nontoxicity established by the U.S. Food and Drug Administration.

Flame retardants are used to hinder ignition or the spread of flames. The most widely used are alumina trihydrate, phosphorus compounds, and halogenated compounds in combination with a synergist such as antimony oxide.

Heat stabilizers are used to prevent degradation of PVC

during processing (PVC is unstable to both heat and light). Stabilizers include organotin mercaptides, methyl and butyl tins, and cadmium/zinc, barium/cadmium, and barium/cadmium/zinc. Not all of these are acceptable in food-packaging applications.

Appendix D
Barrier Resins

Barrier resins protect other packaging materials from absorbing gases, odors, fragrances, and solvents. Development of these resins enabled the use of plastics for products previously packaged in glass.

The two most common barrier resins are ethylene vinyl alcohol and polyvinylidene chloride.

Ethylene vinyl alcohol (EVOH) resins are hydrolyzed copolymers of vinyl alcohol and ethylene. The vinyl alcohol base has exceptionally high gas-barrier properties, but is water soluble and difficult to process. Copolymerization with ethylene retains the high gas-barrier properties and improves moisture resistance and processing. EVOH resins can be used in coextruded structures, films, and coatings of various substrates or monolayers. Since EVOH resins have poor adhesion to most polymers, adhesive resins or "tie resins" must be used.

Polyvinylidene chloride (PVDC) copolymers are based on vinylidene chloride polymerized with methyl acrylate, vinyl chloride, or unsaturated carboxylic acids. Typical formulations contain 2–10 percent plasticizers such as dibutyl sebacate or diisobutyl adipate and 0.5–1 percent heat stabilizer. The most significant property of PVDC resins is their barrier to gases and liquids; they also exhibit resistance to fats, oils, and many chemicals. PVDC resins are used in food packaging as barriers to moisture, flavor, odors, and gases. They are also used to coat paper and polypropylene film.

Notes

CHAPTER 1

1. Research: Ray Ching, Plastics Research Institute, Rutgers University, as reported at Association of New Jersey Recyclers meeting, February 16, 1988.

2. T. Randall Curlee, *The Economic Feasibility of Recycling: A Case Study of Plastic Wastes* (New York, Praeger, 1986).

3. Ibid., p. 81.

4. Ibid.

5. Ibid., pp. 73–74.

6. Franklin Associates, for U.S. Environmental Protection Agency, *Characteristics of Municipal Solid Waste in the United States, 1960 to 2000*, 1988 update, pp. 15–16.

7. "Materials '87," *Modern Plastics*, January 1987, pp. 55–65.

8. "Resins '88," *Modern Plastics*, January 1988, pp. 63–105.

9. *Modern Plastics Guide to Plastic Packaging*, ed., Joseph A. Sneller (New York: McGraw-Hill, 1987).

10. Chem Systems, Inc., for the Society of the Plastics Industry, Inc., *Plastics: A.D. 2000*, July 1987, p. 29.

11. Ibid., p. 26.

12. Curlee, *Economic Feasibility*, p. 166.

13. Ibid., p. 87.

14. Ibid., p. 74.

15. Ibid., p. 15.

16. Ibid., p. 74.

17. Ibid.

18. Chem Systems, *Plastics: A.D. 2000*, p. 38.

19. Curlee, *Economic Feasibility*, p. 74, pp. 166–170.

20. Ibid.

21. Ibid.

22. Ibid.

23. "Special Report: Making Sense of the Materials Explosion," *Plastics Technology*, June 1987, p. 67.

24. "Plastics' Hottest Markets," *Plastics World*, January 1987, p. 31.

25. Resource Integration Systems, *Statewide Market Study for Recyclable Plastics*, Michigan Department of Natural Resources, February 1987, p. 17.

26. Curlee, *Economic Feasibility*, p. 74.

27. Ibid., pp. 167–171.

28. Chem Systems, *Plastics: A.D. 2000*, p. 22.

29. Ibid., p. 23.

CHAPTER 2

1. Franklin Associates, *Characteristics of Municipal Solid Waste*, 1988 update, p. 16.

2. Franklin Associates, *Characteristics of Municipal Solid Waste*, 1986.

3. Christopher C. Lai, Susan E. Selke, and David I. Johnson, "Impact of Plastic Packaging on Solid Waste," Michigan State University, February 1987.

4. Joseph Bruno, *Incentives for Recycling*, New York Legislative Commission on Solid Waste, January 1988, pp. 23–27.

5. Ronald Balazik and Barry W. Klein, *The Impact of Advanced Materials on Conventional Nonfuel Mineral Markets: Selected Forecasts for 1990–2000*, U.S. Department of the Interior, 1987.

6. Resource Integration Systems, *Statewide Market Study*, p. 22.

7. Lai, Selke, and Johnson, "Impact of Plastic Packaging," p. 13.

8. William C. Drennan, "Coping with the Packaging Explosion," *Food Engineering*, September 1987, p. 69.

9. Ibid., p. 71.

10. Sterling, Anthony, "Toward Better Packaging Design Research," *Prepared Foods*, January 1988, p. 111.

11. Personal communication, E. Feldman with Mel Druin, Vice President, Packaging, Campbell's Food Company, June 3, 1988.

12. Personal communication, N. Wolf with Dr. Dan Toner, Senior Research Fellow, Campbell's Food Company, December 7, 1988.

13. Bryan Salvage and Jack Mans, "Top Shelf: A New Product for a New Category," *Prepared Foods* 156 (8), p. 196. August, 1987.

14. Ibid., p. 197.

15. Personal communication, E. Feldman with Patrick VanKeuren, Vice President, Government Affairs, American National Can Company, June 3, 1988.

16. Personal communication, N. Wolf with Eugene Goodenough, Director of Operations Support, Best Foods, March 1989.

17. Ibid.

18. "Plastics' Hottest Markets," p. 33.

19. Rod Bailey: "The Environmental Impact of Plastic Shopping

Bags," Monroe County (N.Y.) Environmental Council – Committee on Plastic Shopping Bags, February 1987, p. 6.

20. Personal communication, E. Feldman with Mary Ellen Gowin, Vice President of Consumer Affairs, Shop-Rite Stores, April 7, 1988.

21. Midwest Research Institute, *Resource and Environmental Profile Analysis of Polyethylene and Kraft Paper Grocery Sacks*, April 1980.

22. Case and Company, "Report of Paper vs. Plastic Bag Handling Costs," September 1984.

23. Midwest Research Institute, *Environmental Impacts of Polystyrene Foam and Molded Pulp Meat Trays*, April 1972.

24. Personal communication, N. Wolf with Bruce Sweyd, Vice President, Original New York Seltzer, October 1988.

25. Franklin Associates, for the Aluminum Association, "The Fate of Used Beverage Containers in the State of New York – Summary," July 1986.

26. Personal communication, N. Wolf with Roger Friskie of Amoco Foam, February 1989.

27. *Plastics World*, January 1987.

28. Personal communication, E. Feldman with representatives of Yo-Plait Company, June 1988.

29. Personal communication, N. Wolf with representatives of The Society of the Plastics Industry, spring 1984.

30. Personal communication, N. Wolf with Stephen Gallagher, Waste Management Director of Environmental Action Coalition, Feburary 1989.

31. Personal communication, E. Feldman with Mary Ellen Gowin, April 1988.

32. Personal communication, N. Wolf with Barbara Levine, CB 2, Brooklyn, and Anthony Guiliani, CB 3, Staten Island, December 1988.

CHAPTER 3

1. Anthony Andrady, "Plastics in the Marine Environment," Proceedings of Symposium on Degradable Plastics, The Society of the Plastics Industry, Washington, D.C., June 10, 1987.

2. Ibid.

3. Representatives of New York City Department of Sanitation, in various reports on landfill closures.

4. William Rathje, Report to National Recycling Congress, St. Paul, Minn., September 1988.

5. Personal communication, E. Feldman with New York City Department of Sanitation representatives at Fresh Kills Landfill, July 1988.

6. Elmer Kaiser and Arrigo Carotti, "Municipal Incineration of Refuse with 2 Percent and 4 Percent Additions of Four Plastics: Polyethylene, Polyurethane, Polystyrene and Polyvinyl Chloride," New York University, 1971.

7. Letter to Dr. Buzz Hoffman, Chief, Environmental Impact Section, Center for Food Safety and Applied Nutrition, Food and Drug Ad-

ministration, from Richard W. Seelinger, Vice President, Ogden Martin Systems, July 2, 1984.

8. Industronics, Inc., "Test Burn Report, PET Bottles," 1983.

9. Midwest Research Institute, for New York State Energy Research and Development Agency, "Results of the Combustion and Emissions Research Project at the Vicon Incinerator Facility in Pittsfield, Massachusetts," June 1987.

10. California Air Resources Board, "Evaluation Test on a Hospital Refuse Incinerator at Saint Agnes Medical Center, Fresno, California," January 1987.

11. William P. Linak, and others, "Waste Characterization and the Generation of Transient Puffs in a Rotary Kiln Incinerator Simulator," Proceedings of the 13th Annual Research Symposium on Land Disposal, Remedial Action, Incineration and Treatment of Hazardous Waste, Environmental Protection Agency, Cincinnati, Ohio, 1987.

12. Marjorie Clarke, "Minimizing Emissions from Resource Recovery," presented at the International Workshop in Municipal Waste Incineration, Montreal, Canada, October 1987, p. 12.

13. Jack Lauber, "Toxic Emissions from Small Incinerators," New York State Department of Environmental Conservation, 1986, p. 3.

14. Clarke, "Minimizing Emissions," p. 15.

15. Charles Papke, "Plastic Recycling as a Business," *Resource Recycling*, September/October 1986, pp. 12–13, 32.

16. Jacob Leidner, *Plastics Waste, Recovery of Economic Value* (New York: Marcel Dekker, Inc., 1981), p. 3.

17. "Resin Suppliers Organize for Solid Waste Battles," *Modern Plastics*, July 1988, pp. 14–16.

18. Robert Leaversuch, "Industry Begins to Face Up to the Crisis of Recycling," *Modern Plastics*, March 1987, p. 45.

19. Society of the Plastics Industry, *Plastic Bottle Recycling Directory and Reference Guide, 1987*, p. 35.

20. Ibid.

21. Correspondence from Jack Bray, Wellman, Inc., January 1988.

22. *Plastic Bottle Recycling Directory.*

23. Correspondence from Jack Bray.

24. Personal communication, E. Feldman with various PET processors, spring 1988.

25. Chem Systems, *Plastics: 2000 A.D.*, p. 70.

26. Ibid.

27. Ibid.

28. Personal communication, E. Feldman with Al Voegele, President, Vermont Recycling Industries, October 1987.

29. Personal communication, E. Feldman with Robert Barrett, Mobil Chemical Company, May 1988.

30. Gary Chamberlain, "Recycling Plastics, Building Blocks of Tomorrow," *Design News*, May 4, 1987, p. 52.

31. Advanced Recycling Technology, Ltd., "The ET/1 Recycling Process," 1988.

32. Ibid.

33. Gretchen Brewer, "State Planning for Post-Consumer Plastics Recycling." Massachusetts Department of Recycling, May 1987, p. 12.

34. "Wet Granulation Cleans Up Film Scrap in Complete Recycling System," *Modern Plastics*, May 1988, p. 17.

35. EAC New York State Recycling Network research, spring 1988 and spring 1989.

36. Personal communication, E. Feldman with Joanne Pappas, Recycling Coordinator, Mt. Olive, New Jersey, May 1988.

37. Ibid.

38. Gretchen Brewer, "Plastics Recycling Action Plan for Massachusetts," Massachusetts Department of Environmental Quality Engineering, Division of Solid Waste Management, July 1988.

39. Personal communication, E. Feldman with Terry Bisser, Rhode Island Solid Waste Management Corporation, June 1988.

40. Janet Keller, "Recovery of Post-Consumer Plastics through Curbside Collection — Two Rhode Island Case Histories," presented at North American Recycling Congress, April 29, 1988.

41. Brewer, "Plan for Massachusetts."

42. Personal communication, N. Wolf with representatives of Center for Plastics Recycling Research, Rutgers University, February 1989.

43. Personal communication, E. Feldman with Luke Schmidt, President, National Association for Plastic Container Recovery, July 1988.

44. Ibid.

45. Personal communication, N. Wolf with Kimberlee Vollbrecht, Northeast Regional Director, Council for Solid Waste Solutions, February 15, 1989, Newark, N.J.

CHAPTER 4

1. Local Ordinance #6S&FA, passed, February 15, 1989, Newark, N.J.

2. Personal communication, N. Wolf with Ronald Forster, Rutgers University, February 1989.

3. Statement of Richard Renfree, Rutgers University, as part of deliberations of Municipal Solid Waste Technology Conference, sponsored by U.S. Environmental Protection Agency, San Diego, Calif., January/February 1989.

4. Rathje, September 1988.

5. Eaglebrook communication, 1989.

6. Personal communication, E. Feldman with Iowa state representatives, December 1987.

7. Will Ferretti and others, Rockefeller Institute at State University of New York at Albany, "Study of New York State's Returnable Container Law for Department of Environmental Conservation," 1984.

8. Franklin Associates, "Fate of Used Beverage Containers."

Bibliography

Advanced Recycling Technology, Ltd. "The ET/1 Recycling Process."
1987.

Agoos, Alice, and Peter Savage. "Riding High, Polystyrene Still Faces a
Triple Threat." *Chemical Week*, December 9, 1987, p. 24.

Andrady, Anthony. "Plastics in the Marine Environment." Proceedings
of Symposium on Degradable Plastics, The Society of the Plastics
Industry, Washington, D.C., June 10, 1987.

Anthony, Sterling. "Toward Better Packaging Design Research." *Prepared
Foods*, January 1988, pp. 111–112.

"As Ozone Evidence Mounts, Industry Cuts Use of CFCs." *Modern Plastics*, May 1988, pp. 10–11.

Bailey, Rod. "The Environmental Impact of Plastic Shopping Bags." Monroe County, (N.Y.) Environmental Council, February 1987.

Balazik, Ronald, and Barry W. Klein. *The Impact of Advanced Materials
on Conventional Nonfuel Mineral Markets: Selected Forecasts for
1990–2000.* United States Department of the Interior, 1987.

Berlaint, Adam. "Man-made Killer of the Seas." *U.S. News and World
Report*, July 6, 1987, p. 72.

Brewer, Gretchen. "State Planning for Post-Consumer Plastics Recycling." Massachusetts Department of Recycling, May 28, 1987.

Brewer, Gretchen. "Plastics Recycling Action Plan for Massachusetts."
Massachusetts Department of Environmental Quality Engineering,
Division of Solid Waste Management, July 1988.

Bruno, Joseph. *Incentives for Recycling.* New York Legislative Commission on Solid Waste Management, January 1988.

California Air Resources Board. "Evaluation Test on a Hospital Refuse
Incinerator at Saint Agnes Medical Center, Fresno, California." January 1987.

Case and Company. "Report of Paper vs. Plastic Bag Handling Costs."
September 1984.

Center for Plastics Recycling Research. "Technical Report #12: Environmental Impact of Plastics Disposal in Municipal Solid Wastes."
Rutgers University, New Brunswick, New Jersey, January 1–June 30,
1986.

Chamberlain, Gary. "Recycled Plastics, Building Blocks of Tomorrow." *Design News*, May 4, 1987.

Chem Systems, Inc. *Plastics: A.D. 2000*. The Society of the Plastics Industry, 1987.

Clarke, Marjorie. "Minimizing Emissions from Resource Recovery." Presented at the International Workshop in Municipal Waste Incineration, Montreal, Canada, October 1–2, 1987.

Clarke, Marjorie. "How Plant Operators Can Minimize Emissions." *Waste Age*, December 1987.

Collaten, Lisa. "Do We Really Need More Plastic?" *Public Citizen*, July/August 1987.

Curlee, T. Randall. *The Economic Feasibility of Recycling: A Case Study of Plastic Wastes*. New York: Praeger, 1986.

Drennan, William C. "Coping with the Packaging Explosion." *Food Engineering*, September 1987.

"EA Reports." *Environmental Action Magazine*, July/August 1988, p. 4.

Ecoplastics, Ltd. "Toxicological Assessment of Leachate from Photodegraded Ecolyte Polyethylene." Summary report by Beak Consultants, Ltd., February 1988.

Encyclopedia of Polymer Science and Engineering. John Wiley and Sons, New York City, 1986.

Englebright, Steve. "Local Law 1869–87, a Local Law to Simplify Solid Waste Management by Requiring Certain Uniform Packaging Practices within the County of Suffolk." Suffolk County, New York, 1987.

Epstein, Eliot, and Todd Williams. "Solid Waste Composting Gains New Credence." *Solid Waste and Power*, June 1988, pp. 24–26.

Ferrand, Trish. "Research in Plastics Recycling." *Resource Recycling*, March/April 1987, pp. 18–19, 48.

"Florida Readies Broad Assault on Garbage." *Wall Street Journal*, July 20, 1988.

Frados, Joel. *The Story of the Plastics Industry*. The Society of the Plastics Industry, May 1977.

Franklin Associates, Ltd. "The Fate of Used Beverage Containers in the State of New York – Summary." The Aluminum Association, July 1986.

Franklin Associates, Ltd. "Characteristics of Municipal Solid Waste in the United States, 1960 to 2000, (1988 Update)." The Environmental Protection Agency, 1988.

Fuller, E. E. "Plastics Packaging Recycling." Presented at the New York State Legislative Commission on Solid Waste Management Conference, New York City, January 28, 1988.

Gottesman, Roy T., and Donald Goodman. "Polyvinyl Chloride." *Applied Polymer Science*. 2d ed. American Chemical Society, 1985.

Halek, George W. "Tutorial Lecture on Plastics." Presented at the Plastics Recycling Symposium at Rutgers University, New Brunswick, New Jersey, March 28, 1985.

Hellman, Eric. *Plastics Recycling: Understanding the Opportunities.* Toronto Recycling Action Committee, 1981.

Industronics, Inc. "Test Burn Report, PET Bottles." 1983.

Kaiser, Elmer, and Arrigo Carotti. "Municipal Incineration of Refuse with 2 Percent and 4 Percent Additions of Four Plastics: Polyethylene, Polyurethane, Polystyrene and Polyvinyl Chloride." New York University, 1971.

Kaiser, Elmer, and Arrigo Carotti. "Incineration of Municipal Refuse with 2% and 4% Additions of Polyethylene Terephthalate." New York University, 1971.

Keller, Janet. "Recovery of Post-Consumer Plastics through Curbside Collection—Two Rhode Island Case Histories." Presentation at North American Recycling Congress, April 29, 1988.

Keller and Heckman Law Offices. Report on Food Contact Usage of PVC Plastics. August 17, 1983.

Kelly, Keith. "Consumers Will Pay for Protecting Ozone." *Newsday,* April 12, 1988.

Lai, Christopher C., Susan E. Selke, and David I. Johnson. "Impact of Plastic Packaging on Solid Waste." Michigan State University, East Lansing, February 1987.

Lauber, Jack. "Toxic Emissions from Small Incinerators." New York State Department of Environmental Conservation, New York City, 1986.

Leaversuch, Robert. "Industry Begins to Face Up to the Crisis of Recycling." *Modern Plastics,* March 1987.

Leaversuch, Robert. "Is Industry Serious about Solid Waste Recovery?" *Modern Plastics,* June 1988, pp. 65–76.

Leidner, Jacob. *Plastics Waste, Recovery of Economic Value.* New York: Marcel Dekker, Inc., 1981.

Levy, Michael. "A Plastics Packaging Perspective on Waste Reduction." Presented at the New York State Legislative Commission on Solid Waste Management Conference, New York City, January 28, 1988.

Linak, William P., and others. "Waste Characterization and the Generation of Transient Puffs in a Rotary Kiln Incinerator Simulator." Proceedings of the 13th Annual Research Symposium on Land Disposal, Remedial Action, Incineration and Treatment of Hazardous Waste, Environmental Protection Agency, Cincinnati, Ohio, 1987.

Lloyd, D. Roger. "Degradable Polymers." Presented at Recyclinplas II Conference, Plastics Institute of America, Washington, D.C., June 18–19, 1987.

"Materials '87." *Modern Plastics,* January 1987.

"Materials '88." *Modern Plastics,* January 1988.

"McDonald's Pullout: CFC Issue Hits Home." *Modern Plastics,* October 1987, pp. 15–16.

Midwest Research Institute. *Environmental Impacts of Polystyrene Foam and Molded Pulp Meat Trays.* April 1972.

Midwest Research Institute. *Resource and Environemntal Profile Anal-*

ysis of Polyethylene and Kraft Paper Grocery Sacks. April 1980.

Midwest Research Institute. "Results of the Combustion and Emissions Research Project at the Vicon Incinerator Facility in Pittsfield, Massachusetts." New York State Energy Research and Development Authority, June 1987.

Mobil Chemical Company. "Polystyrene Foam and Chlorofluorocarbons." 1987.

Modern Plastics Encyclopedia, 1986–1987. Edited by Robert Martino. New York: McGraw-Hill, 1986.

Modern Plastics Encyclopedia, 1988. Edited by Rosalind Jurand. New York: McGraw-Hill, 1988.

Modern Plastics Guide to Plastic Packaging. Edited by Joseph A. Sneller. New York: McGraw-Hill, 1987.

Morrow, Darrell R., and others. "Overview of Plastics Recycling." *Converting and Packaging,* December 1987, p. 139.

New Jersey Alliance for Action and New Jersey Institute of Technology. *Solving the Garbage Crisis in New Jersey: A Cooperative Study.* 1986.

New York State Department of Environmental Conservation. "Results of NYSDEC Clean Ocean Campaign, National Beach Cleanup Day." 1987.

Papke, Charles. "Plastic Recycling as a Business." *Resource Recycling,* September/October 1986, pp. 12–13, 32.

Pess, George. *Thermoplastics in the Post-Consumer Waste Stream.* New York, New York Environmental Action Coalition, 1985.

The Plastic Bottle Institute. "Energy Use and Resource Recovery." The Society of the Plastics Industry, 1981.

"Plastic Containers Coded for Recycling." *COPPE Quarterly* (Council on Plastics and Packaging in the Environment), vol. 2, no. 2, spring 1988.

"Plastics' Hottest Markets." *Plastics World,* January 1987.

Portnoy, Kristine. "From Cornstarch, a Biodegradable Film." *Chemical Week,* May 27, 1987, p. 36.

"Proposed Uses of Vinyl Chloride Polymers." *Federal Register,* vol. 51, no. 22, Proposed Rules, February 3, 1986.

Redpath, A. "Photodegradable Controlled Lifetime Plastics: A Strategic Environmental Advantage for the Plastics Industry." Ecoplastics, Ltd., June 10, 1987.

Reinhold, Robert. "California Recycling Plan Is in Jeopardy." *New York Times,* July 4, 1988, p. 7.

Report of the Nelson A. Rockefeller Institute of Government to the Temporary State Commission on Returnable Beverage Containers. "The New York Returnable Beverage Container Law: The First Year." March 1985.

"Resin Suppliers Organize for Solid Waste Battles." *Modern Plastics,* July 1988, pp. 14–16.

"Resins '88." *Modern Plastics,* January 1988.

Resource Integration Systems. *Statewide Market Study for Recyclable Plastics.* Michigan Department of Natural Resources, 1987.

Roosevelt, Mark. "HB 1172, An Act to Protect the Environment by Encouraging a Reduction and Recycling of Packaging in the Commonwealth." Massachusetts House of Representatives, 1987.

Russo, John. "An Act Concerning the Taxation of Certain Plastic Materials and Items, and Supplementing Title 13 of the Revised Statutes." Senate No. 126, State of New Jersey, 1988.

Sacharow, Stanley. "EVOH and PVDC: Rivals or Allies?" *Prepared Foods,* February 1988, pp. 82–83.

Salvage, Bryan, and Jack Mans. "Top Shelf: A New Product for a New Category." *Prepared Foods,* August 1987, pp. 196–198.

Shabecoff, Philip. "DuPont to Halt Chemicals that Peril Ozone." *New York Times,* March 25, 1988, pp. A1, A20.

Smith, N. Dean. Letter to Maggie Clark, November 23, 1987.

Smoluk, George. "Today It's Scrap, Tomorrow It's Feedstock." *Modern Plastics,* May 1988, pp. 48–50.

Snow, Darlene. "Plastics and Other Packaging Under Attack." *Waste Age,* July 1988, p. 133.

Society of the Plastics Industry. *Plastic Bottle Recycling Directory and Reference Guide,* 1987.

"Solving the Problem of Degradability: An Imperative for the Industry." *Modern Plastics,* May 1988, pp. 148–149.

"Special Report: Making Sense of the Materials Explosion." *Plastics Technology,* June 1987.

Tacito, Louis. "Turning Recycled Plastics into Usable Products." Presented at How to Survive Mandatory Recycling and Disposability, Society of Packaging Engineers, November 4, 1987.

United States Environmental Protection Agency. "Fact Sheet – Municipal Waste Combustion Ash." March 1988.

United States Food and Drug Administration. *Environmental Assessment and Finding of No Significant Impact for FDA-Initiated Action on Vinyl Chloride Polymers.* September 16, 1985.

van der Laan, Corrie. "PVC as Food Contact Material." Environmental Task Force, Washington, D.C., December 1987.

VandenBerg, Nancy. "Buying Recycled Products: Barriers to Overcome." Presented at Recyclinplas II Conference, Plastics Institute of America, Washington, D.C., June 18–19, 1987.

Weiss, Samuel. *Sanitary Landfill Technology.* Parkridge, New Jersey: Noyes Data Corporation, 1974.

"Wet Granulation Cleans Up Film Scrap in Complete Recycling System." *Modern Plastics,* May 1988, p. 17.

Wicks, John. "PVC: The Facts." Association of Plastics Manufacturers of Europe, 1986.

Index

124

About the Authors

NANCY WOLF, Executive Director of the Environmental Action Coalition, has been actively engaged in the environmental field for the past twenty years. Her work has included the development of written and audio-visual materials for teachers and students. As an advisor to New York City and New York State governments, she has served on numerous commissions, particularly those pertaining to solid-waste management.

Ms. Wolf is the mother of two daughters and lives with her husband in New York City. She is a graduate of Hollins College and gained her master's degree at The Johns Hopkins University.

ELLEN FELDMAN, formerly Science Associate for the Environmental Action Coalition, is presently an environmental planning consultant with Eldon Environmental Services in Great Neck, New York. Her current work focuses on the compositing of solid waste with special attention to plastics.

Ms. Feldman is a graduate of the University of Pennsylvania and received her master's degree from Rutgers University. She lives with her husband in Long Island, New York.

Also Available from Island Press

Ancient Forests of the Pacific Northwest
By Elliott A. Norse

Better Trout Habitat: A Guide to Stream Restoration and Management
By Christopher J. Hunter

The Challenge of Global Warming
Edited by Dean Edwin Abrahamson

Coastal Alert: Ecosystems, Energy, and Offshore Oil Drilling
By Dwight Holing

The Complete Guide to Environmental Careers
The CEIP Fund

Creating Successful Communities: A Guidebook for Growth Management Strategies
By Michael A. Mantell, Stephen F. Harper, and Luther Propst

Crossroads: Environmental Priorities for the Future
Edited by Peter Borrelli

Economics of Protected Areas
By John A. Dixon and Paul B. Sherman

Environmental Restoration: Science and Strategies for Restoring the Earth
Edited by John J. Berger

Fighting Toxics: A Manual for Protecting Your Family, Community, and Workplace
By Gary Cohen and John O'Connor

Hazardous Waste from Small Quantity Generators
By Seymour I. Schwartz and Wendy B. Pratt

Holistic Resource Management Workbook
By Alan Savory

In Praise of Nature
Edited and with essays by Stephanie Mills

Natural Resources for the 21st Century
Edited by R. Neil Sampson and Dwight Hair

The New York Environment Book
By Eric A. Goldstein and Mark A. Izeman

Overtapped Oasis: Reform or Revolution for Western Water
By Marc Reisner and Sarah Bates

Permaculture: A Practical Guide for a Sustainable Future
By Bill Mollison

The Poisoned Well: New Strategies for Groundwater Protection
Edited by Eric Jorgensen

Race to Save the Tropics: Ecology and Economics for a Sustainable Future
Edited by Robert Goodland

Recycling and Incineration: Evaluating the Choices
By Richard A. Denison and John Ruston

Resource Guide for Creating Successful Communities
By Michael A. Mantell, Stephen F. Harper, and Luther Propst

Rivers at Risk: The Concerned Citizen's Guide to Hydropower
By John D. Echeverria, Pope Barrow, and Richard Roos-Collins

Rush to Burn: Solving America's Garbage Crisis?
From *Newsday*

Shading Our Cities: A Resource Guide for Urban and Community Forests
Edited by Gary Moll and Sara Ebenreck

War on Waste: Can America Win Its Battle With Garbage?
By Louis Blumberg and Robert Gottlieb

Wetland Creation and Restoration: The Status of the Science
Edited by Mary E. Kentula and Jon A. Kusler

Wildlife and Habitats in Managed Landscapes
Edited by Jon E. Rodiek and Eric G. Bolen

Wildlife of the Florida Keys: A Natural History
By James D. Lazell, Jr.

For a complete catalog of Island Press publications, please write:
Island Press
Box 7
Covelo, CA 95428

or call: 1-800-828-1302

Island Press Board of Directors